Hazardous Waste Tracking and Cost Accounting Practice

Hazardous Waste Tracking and Cost Accounting Practice

Richard T. Enander

CRC Press
Taylor & Francis Group
Boca Raton London New York

CRC Press is an imprint of the
Taylor & Francis Group, an **informa** business

First published 1996 by Lewis Publishers

Published 2021 by CRC Press
Taylor & Francis Group
6000 Broken Sound Parkway NW, Suite 300
Boca Raton, FL 33487-2742

© 1996 by Taylor & Francis Group, LLC
CRC Press is an imprint of Taylor & Francis Group, an Informa business

No claim to original U.S. Government works

ISBN 13: 978-1-56670-142-6 (hbk)

DOI: 10.1201/9780138752234

Visit the Taylor & Francis Web site at
http://www.taylorandfrancis.com

and the CRC Press Web site at
http//www.crcpress.com

Library of Congress Cataloging-in-Publication Data

Enander, Richard T.
   Hazardous waste tracking and cost accounting practice / Richard T.
   Enander
      p.   cm.
   Includes bibliographical references and index.
   ISBN 1-56670-142-2
   1. Hazardous wastes--Tracking.   2. Cost accounting.   I. Title.
TD1060.T73E59 1995
828.4'2'068--dc20   94-41779
                                                                           CIP
Library of Congress Card Number 94-41779

This book was written by Richard T. Enander in his private capacity. No official support or endorsement by the Rhode Island Department of Environmental Management or any other agency of the State or Federal Government is intended or should be inferred.

To my wife Rebeca and my son Richard Anthony
with whom I am truly blessed — no husband
or father could ask for more

AND

To my mother Marie for her love and devotion
that only a mother could provide, and my father Richard Lans
from whom I have learned patience, hard work and perseverance,
and to my sister Deborah (and family Len, Melissa and Stan) who
is always there in time of need and with whom I have shared
some of the more important moments in life.

Richard T. Enander is a Principal Environmental Scientist and Pollution Prevention Program Manager with the Rhode Island Department of Environmental Management's Office of Environmental Coordination, Providence. In this capacity he leads a team of engineers, scientists and planners working to reduce pollution in a wide range of industries including auto-refinishing, textile manufacturing, metal finishing, chemical manufacturing and seafood processing. He formerly worked as an environmental specialist for the Specialty Chemicals Group of Hoechst Celanese Corporation — a subsidiary of Hoechst AG, one of the largest chemical companies in the world. Mr. Enander holds B.S. and M.S. degrees in Environmental Science and Environmental Health, respectively, with concentrations in chemistry, biology and environmental engineering. He has authored or co-authored numerous articles on waste management/minimization and pollution prevention. He is a U.S. EPA Fellow, a member of the American Chemical Society and is currently pursuing a Ph.D. in Environmental Health at Tufts University, College of Engineering, where his current research interests are in the areas of exposure and risk assessment.

Preface

As we approach the 21st century, environmental challenges have never been greater. Never before in history have we experienced the need for such diligence in our efforts to track the use, manufacture, processing, treatment, and disposal of toxic and hazardous materials. Legislation passed over the last twenty-five years has not only resulted in improved environmental quality, but has also created new levels of accountability for today's environmental professional.

With the passage of each new environmental law — specif. RCRA (1976), CERCLA (1980), HSWA (1984), EPCRA (1986) and the Pollution Prevention Act (1990) — the need for better methods of tracking hazardous waste management data at the facility level became increasingly self-evident. By the mid-to-late 1980s, the integration of a cost accounting function into point of generation waste tracking systems (i.e., manual/electronic data management systems that track wastes from the generating source to final treatment and/or disposal) began to emerge as an important step in the overall design of corporate environmental management programs.

Hazardous Waste Tracking and Cost Accounting Practice is written in response to this progressive movement toward improved waste material tracking and accounting systems. It describes the fundamental principles of hazardous waste tracking by point of generation and "fully-loaded" waste management cost accounting. The intent of this work is not to provide an academic methods review, but rather to focus on those objectives that are practical and readily attainable. Toward this end, specific guidance is given on how to establish recordkeeping and accounting systems that are both effective and broadly adaptable in application. The organizational framework presented herein can be tailored to meet individual site-specific needs or used as a point of reference in the design of alternate systems.

Though written for the practicing professional who is compelled to keep facility records relating to the generation and management of solid and hazardous waste, it is hoped that others engaged in the hazardous waste management/minimization and pollution prevention fields will find this book to be a useful reference.

HOW THIS BOOK IS ORGANIZED

Hazardous Waste Tracking and Cost Accounting Practice is organized into six chapters which contain four basic tracking and cost accounting steps. Each step follows a logical pattern of progression guiding the reader through a series of program building tasks. The first three steps establish the identity, source and properties of each waste. In the fourth and final step, a simple method is described for tracking the movement and "fully-loaded" management costs of each waste from the point of generation to final treatment and/or disposal.

In Chapter 1, a brief historical review of selected federal environmental laws and guidance is presented. Collectively, these laws — while created to address specific environmental problems — have served to drive the need for better hazardous waste tracking and cost accounting in business and industry.

Chapter 2 discusses the general concept of waste material tracking and cost accounting by point of generation. The waste management cycle and its relationship to unit tracking and the development of waste identification systems are reviewed. The organization of data by generating source and the adoption of specific container/bulk material tracking practices are presented as a necessary first step in the development of a comprehensive waste tracking and cost accounting program.

In Chapter 3, the importance of understanding waste generating processes/operations is discussed with reference to specific examples that highlight the complexities of process-based waste determinations. Documenting precisely "how" and "where" wastes are generated is described in this chapter as a key step in waste material tracking. The material balance concept is reviewed and detailed guidance provided on the collection and organization of process-specific data.

Chapter 4 builds upon the information gathered in Ch. 3 and focuses on the development of physical, chemical, health and regulatory profiles for each waste. The use of data sheets in generator tracking and hazardous waste management programs is reviewed. Tracking techniques for chemical mixtures, in support of waste determinations based on generator knowledge, are reviewed and compared to laboratory testing as the basis for waste characterization. The performance of waste hazard analyses and communication of waste material hazards to workers are also presented as important components of this third step.

Chapter 5 provides an overview of shipment and cost data needs relative to off site activity tracking and "fully-loaded" waste management cost accounting. The three general types of data and information (i.e., manifest and point of generation data; treatment, storage and disposal facility final disposition reports; and, complete management cost data) that are needed to track wastes and costs by point of generation are described in detail.

The fourth and final waste material tracking and cost accounting step is presented in Chapter 6. In this chapter, guidance is given on how to organize tracking and cost data into a rather simple spreadsheet accounting system. The method presented provides a basis for (1) tracking each unit of waste and multiple sets of interrelated data across the entire management cycle, and (2) allocating "fully-loaded" waste management costs directly back to the product, operation or department of origin. Fundamental statistical tools that can be used as an aid in the analysis and presentation of waste management data are also reviewed in this chapter.

Richard T. Enander
West Greenwich, Rhode Island

Acknowledgments

I am indebted to the many individuals who over the years have shared their time, ideas and experiences which ultimately led to the development of this book. I would especially like to thank Antonio Ramírez for his help during the initial design of the recordkeeping forms presented herein; Brígida Yolanda Montúfar for her assistance during the final hours of manuscript preparation; Donald J. Nester who introduced me to the concept of environmental problem-solving through employee involvement; Dr. Eugene Park for his comments concerning flow diagram construction; Peter W. Egan, a colleague who provided materials and exhibits; Adjunct Asst. Prof. Anne Marie C. Desmarais, Tufts University, who through her lectures and critical evaluations provided valuable instruction in toxicology and toxicity profile development; and Dr. David M. Gute, Associate Prof., Tufts University, for his support in my pursuit of advanced graduate training which has helped me to create a more complete text.

Contents

List of Figures

List of Tables

Chapter 1

HISTORICAL PERSPECTIVE ON REGULATORY INITIATIVES DRIVING HAZARDOUS WASTE TRACKING AND COST ACCOUNTING

The concept of waste tracking is not new, but the degree of administrative, technical and regulatory involvement is. In the past, U.S. industry recorded waste movement by relatively simple material tracking means and often for the principal purpose of accounting for off site transportation and disposal costs. Since the birth of the "cradle-to-grave" system on October 21, 1976, however, solid and hazardous waste tracking has never been the same. On that day, the Resource Conservation and Recovery Act (RCRA) was enacted by the 94th Congress of the United States as an amendment to the Solid Waste Disposal Act of 1965. This milestone marked the creation of a new era in waste material tracking, environmental protectionism and corporate accountability. For the first time, a subset of solid wastes would be subject to national hazardous management standards and systematically tracked from the "site" of generation to off site treatment, storage and disposal.

CERCLA

On December 11, 1980, Congress passed the Comprehensive Environmental Response, Compensation and Liability Act — also known as CERCLA or Superfund. Complementing RCRA's management and control program, this new statute focused national attention on releases of hazardous substances to the environment. In addition to structuring a funding mechanism and federal framework for hazardous substance response, removal and remedial actions, Superfund put U.S. industry on notice of the importance of proper hazardous substance handling and waste management. The ultimate disposition and fate of all hazardous substances and regulated wastes quickly became central issues in corporate decisionmaking.

HSWA AND EPCRA

In 1984, the scope of RCRA was broadened by the Hazardous and Solid Waste Amendments (HSWA). The new RCRA contained new requirements for waste minimization certification and reporting, expanded the

1

number of federally regulated generators by nearly twelvefold and embraced more than seventy new provisions. Waste minimization (i.e., the reduction of hazardous waste at the source and recycling for beneficial use/ reuse or materials reclamation) became a new national priority calling for companies to focus their efforts on reducing the generation of hazardous waste, while also improving methods to track each unit of waste from the point of generation to final treatment and disposal.

Two years later, on October 17, 1986, CERCLA was also amended. These amendments, embodied in the Emergency Planning and Community Right-To-Know Act (EPCRA), called for annual reporting of toxic chemical releases and represented an unprecedented national push toward comprehensive emergency planning, information collection and dissemination. As a result, many environmental managers found themselves, for the first time, performing "material balance" calculations in an effort to quantify off site transfers and toxic chemical releases to the environment. From the mid-1980s onward, hazardous materials and waste tracking began to emerge as an important component of environmental management systems.

EPA'S GUIDANCE TO GENERATORS FOCUSES ON WASTE TRACKING AND COST ALLOCATION

With much of the command and control legislation firmly in place, the U.S. Environmental Protection Agency (EPA) soon began to focus more and more of its attention on reducing or eliminating the generation of wastes at their source. On January 26, 1989, EPA released its first official Pollution Prevention Policy Statement[1] and less than five months later issued draft guidance to generators on six elements that an "effective waste minimization program should include."[2] One of the key elements identified by EPA called for the establishment of "cost allocation systems" whereby "departments and managers are charged fully-loaded waste management costs* for the wastes they generate, factoring in liability, compliance and oversight costs."

In May 1993, EPA updated the 1989 guidance and continued to emphasize the importance of waste tracking and cost accounting. In this latest version, EPA recommended that company business decisions take into account the true costs associated with hazardous materials management. Understanding these costs was described as a necessary first step if a busi-

---

* The term "fully-loaded waste management costs" is defined in this text as the sum total of all on site and off site costs associated with the generation and management of solid and hazardous waste. These costs may include: packaging materials, equipment time, storage space, administrative costs, in-plant labor, lab packing/repackaging service costs, waste analysis costs, future liabilities, transportation/demurrage fees, taxes, surcharges and off site treatment, storage and disposal costs.

ness enterprise is to make an informed decision on the potential savings associated with waste minimization. EPA suggested that these costs should include the cost of internal regulatory oversight, paperwork and reporting requirements, loss of production potential, transportation, off site storage, treatment and disposal costs, employee exposure, liability insurance and future liability (e.g., CERCLA).[3]

EPA's vision of what constitutes a practical and effective cost allocation system is described as having the following features:[4]

> [Companies should] maintain a waste accounting system to track the types and amounts of wastes as well as the types and amounts of the hazardous constituents in wastes, including the rates and dates they are generated. EPA realizes that the precise business framework of each waste generator can be unique. Therefore, each organization must decide the best method to obtain the necessary information to characterize waste generation. Many organizations track their waste production by a variety of means and then normalize the results to account for variations in production rates.

> Additionally, a waste generator should determine the true costs associated with waste management and cleanup, including the costs of regulatory compliance, paperwork and reporting requirements, loss of production potential, cost of materials found in the waste stream (perhaps based on the purchase price of those materials), transportation/treatment/storage/disposal costs, employee exposure and health care, liability insurance, and possible future RCRA or Superfund corrective action costs. Both volume and toxicities of generated hazardous waste should be taken into account. Substantial uncertainty in calculating many of these costs, especially future liability, may exist. Therefore, each organization should find the best method to account for the true costs of waste management and cleanup.

> Where practical and implementable, organizations should appropriately allocate the true costs of waste management to the activities responsible for generating the waste in the first place (e.g., identifying specific operations that generate the waste rather than charging the waste management costs to "overhead"). Cost allocation can properly highlight the parts of the organization where the greatest opportunities for waste minimization exist; without allocating costs, waste minimization opportunities can be obscured by accounting practices that do not clearly identify the activities generating the hazardous wastes.

In 1992, EPA published additional cost accounting guidance in its *Facility Pollution Prevention Guide*. This guidebook describes the basic components of a total cost assessment approach to pollution prevention projects. A more detailed discussion of cost accounting, and its relationship to pollution prevention, is provided in Chapters 5 and 6. [Note: Pollution prevention has been defined by the U.S. EPA as the use of materials, processes, or practices that reduce or eliminate the creation of pollutants or wastes at the source (source reduction).[5] It is broader in meaning and application than the term *waste minimization* in that it (1) includes energy and water conservation, and (2) applies to pollutants of all types, not just hazardous waste.]

A NEW DIRECTION

Tracking and assigning fully-loaded costs to waste generating processes was, and continues to be, seen as a way to encourage reductions in the use and disposal of toxic and hazardous materials. The national movement away from waste management as the dominant environmental strategy (and toward source reduction) was further strengthened and took on new proportions with the passage of the Pollution Prevention Act on October 27, 1990.

Over a 17 year period, from 1976 to 1993 (and continuing to the present), a clear trend toward improved toxic chemical use and waste management reporting, with an increased emphasis on source reduction, corporate accountability and cost accounting, has emerged. To better meet these and other environmental challenges, it has become clear that recordkeeping and accounting programs must reach new levels of efficiency. The steps presented in this book represent one way to organize administrative efforts in response to these ever-growing data management demands. The materials presented herein are dedicated to the achievement of a more efficient and effective means of program administration, waste material tracking and fully-loaded waste management cost accounting.

REFERENCES

1. **Federal Register**, 54, 3845, January 26, 1989.
2. **Federal Register**, 54, 25056, June 12, 1989.
3. **Egan, P.W., Enander, R.T. and Gouchoe, S.,** EPA's Guidance on Waste Minimzation Outlines Program Elements, in *Journal of Environmental Regulation*, John Wiley & Sons, Inc., 605 Third Avenue, New York, N.Y. 10158-0012. August 1994.
4. **Federal Register**, 58, 31114, May 28, 1993.
5. **U.S. Environmental Protection Agency,** *Facility Pollution Prevention Guide*, Office of Research and Development, Washington, D.C. 20460. May 1992.

Chapter 2

TRACKING BY POINT OF GENERATION
In-Plant Waste Tracking Systems and Inventory Development

INTRODUCTION

The first national push toward hazardous waste tracking came in 1980, when the U.S. Environmental Protection Agency (EPA) introduced its "cradle-to-grave" manifest tracking system. Created in cooperation with the U.S. Department of Transportation (DOT), the manifest system tracks hazardous waste shipments from the generator's "site" boundary to final destination. The Uniform Hazardous Waste Manifest lies at the heart of this system and is the shipping document now used by all generators who "transport, or offer for transportation, hazardous waste for off site treatment, storage or disposal."[1]

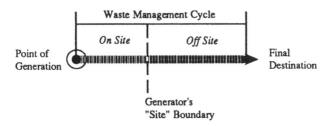

Figure 1. Waste management cycle for off site shipments.

Though the uniform manifest remains the cornerstone of all off site hazardous waste tracking initiatives, it is but one tool currently used by generators to track wastes from their "point of generation" to final treatment and disposal. Consider Figure 1, for example. In this illustration, the entire waste management* cycle is depicted and can be described as follows: As a waste exits a process (point of generation) or is otherwise generated, it must first be collected, then stored and prepared for shipment (on site waste management). On the day of the shipment, a manifest is initiated and serves

---

* Waste management, as defined by EPA, is the systematic control of the collection, source separation, storage, transportation, processing, treatment, recovery, and disposal of hazardous waste.[2]

to track aggregate quantities of waste and off site waste management activities from the generator's "site" boundary to final destination.

From an administrative perspective, tracking aggregate quantities of waste from the generator's site boundary to a designated facility* is a relatively simple task. Proper completion of the uniform manifest, distribution and receipt of return copies is all that's required. Once a waste shipment leaves the generator's property, the manifest system is set into motion and plays a key role in off site waste material tracking.

By comparison, tracking each unique waste and complete management costs by "point of generation" to their ultimate or final destination can prove to be more difficult, especially when wastes are shipped to intermediate storage facilities. When a company ships to an intermediate storage facility or when wastes are generated at many different locations within a single plant, the manifest system must be supported by additional recordkeeping methods that can track each unit of waste by "point of generation" across the entire waste management cycle. The guidance provided in this chapter will help you to develop a point of generation tracking system that meets your professional needs. This system will then form the basis for your continued efforts in the steps presented in Chapters 3, 4, 5 and 6.

GETTING STARTED

BENEFITS TO TRACKING
WASTES BY POINT OF GENERATION

As wastes move through the management cycle, generators can sometimes lose track of their identity, history of origin, final disposition or true management costs. This can occur at several points in the management cycle and is a problem that is common to companies both large and small. One way to overcome these and other potential problems is to develop a recordkeeping system that tracks and organizes key hazardous waste and cost accounting data by point of generation. This is a very simple process that can yield unexpected benefits with only a modest investment in time.

Some benefits to tracking hazardous waste and fully-loaded waste management cost data by point of generation include:

• *Responding to Environmental Challenges* — Laws, regulations and cooperative initiatives (for example, state/federal pollution prevention

---

* Designated facility means a hazardous waste treatment, storage, or disposal facility which has received an EPA permit and has been designated on the manifest by the generator.[3] [When the designated facility is an intermediate storage facility, the entry on the manifest does not represent the final destination of the waste.]

grant/assistance projects and EPA's 33/50 Project) are increasingly focusing on the processes that generate waste. Company programs that track wastes by point of generation have a clear advantage when it comes to planning for and responding to new and existing environmental challenges.

• *Cost Allocation* — When tracked by generating source, complete or "fully-loaded" waste management costs can be allocated directly back to the product, manufacturing process, operation or department of origin. Allocating costs in this way provides departmental managers with a strong incentive to reduce waste, thereby decreasing a company's liability exposure, management costs and overall administrative burden.

• *Trend Analysis* — Changes in (1) the types and characteristics of wastes generated, (2) the rates of waste generation, (3) the methods used to manage wastes, and (4) associated waste management costs are more easily identified and monitored when hazardous waste data are tracked by point of generation.

• *Effective Management* — Improvements in program control and effectiveness can be realized when waste management and cost data are organized and systematically tracked by point of generation across each step in the management cycle. When data are tracked by generating source, overall program efficiency improves as the total time required to generate waste management reports, perform material balance studies or conduct safety, health and environmental reviews is reduced. Also, management problems that are sometimes experienced with off-specification wastes (i.e., wastes that exhibit unusual properties or contain unexpected amounts of certain elements, chemicals, or chemical compounds) or with "orphan" drums can be avoided or, at least, more easily resolved when procedures are in place to track wastes and related data by point of generation.

CHAPTER OVERVIEW

Your first task in developing a comprehensive program is to determine an appropriate system for identifying wastes by point of generation. The overall approach for doing this is rather simple and is typically a function of the size and distribution of a generator's solid and hazardous waste universe. In companies where only a few waste streams are generated, managers generally do not need to employ formal or highly structured methods for identifying wastes by generating source; see Waste Identification Systems, page 8. For these companies, the point of generation may never be an issue and efforts can be more appropriately focused on defining waste material composition, final disposition tracking and full cost accounting (Chapters 3 through 6).

Table 1
Grouping of Tracking System Development Steps in Chapter 2

Developing a Tracking Method:	• The Waste Identification System
	• ID Numbers: Function & Use
	• System Development
Defining the Universe of Wastes:	• Creating an Inventory
Program Implementation:	• Documentation
	• Tracking by ID Number

In companies where wastes are generated at many different locations, however, the waste identification system can be very important to the overall integrity and success of waste material tracking efforts. In these companies, the waste identification system typically takes the form of a series of numbers or codes that are assigned to each unique waste at the time it is first generated. These numbers are then used to index data and to track the movement of each waste, and fully-loaded management costs, across the entire management cycle.

This chapter describes the development and use of a simple method for tracking wastes by point of generation. Sections within this chapter refer to the design of an in-plant tracking system and can be grouped into several broad categories as shown in Table 1. If you are a small generator with relatively few waste streams, much of the discussion that follows may be of only passing interest. Sections entitled System Development, Creating an Inventory and Documentation are, however, applicable to most generators, regardless of size.

DEVELOPING A TRACKING METHOD

WASTE IDENTIFICATION SYSTEMS

The term *waste identification system* is generally used to refer to a standard method for codifying wastes by point of generation. At the heart of this system is the waste identification (ID) number. The waste ID number (or source identifier) is the code assigned to a waste material that identifies its place of origin. It is a control number. This number facilitates recordkeeping and reporting efforts and is systematically assigned to each waste at the time it is first generated. In small companies with relatively few waste streams, the source identifier may not be a number at all, but rather a short title or key word that refers to the type of waste, product or generating process; e.g., used lube oil or degreasing solvent.

Table 2
Waste Codes Assigned to Generators' Waste by
Treatment, Storage and Disposal Facilities [4]

Sample Code	Code Name	Treatment, Storage, Disposal Facility
U26261	Profile Sheet No.	Clean Harbors Inc. (MA)
225004	Waste Material Data Sheet No.	ENSCO, Inc. (AR)
L-4526	RES Stream No.	Rollins Environmental Services Inc. (NJ)
BH4898	Profile No.	Chemical Waste Management, Inc. (IL)
ST-00009-9542	Authorization No.	ThermalKEM (SC)
PW#01286-1101*	Authorization No.	Laidlaw Environmental Services (SC)

* PW# 01286 represents the *generator specific location*; in this case Reidsville, NC. "This 5 digit number does not change. The last 4 digits of the code, 1101, is waste stream specific. Each time the generator permits a new waste through Laidlaw, the last 4 digits of the code will change to identify the specific waste stream." [5] The South Carolina Department of Health and Environmental Control also uses this code for instate waste tracking.

Commercial Applications

Standard methods for identifying wastes by generator "site" have been in use for some time. Most of today's commercial treatment, storage and disposal facilities (TSDFs), for example, track incoming shipments and subsequent management activities by using numerical codes. As an initial step in the approval process, these TSDFs assign codes, sometimes referred to as Profile Sheet, Authorization, or Stream Numbers, to each unique type of waste (examples provided in Table 2). These codes, used to facilitate material tracking, typically appear on waste data sheets, sample containers, invoices and generator contracts. TSDFs often request that generators mark the top (and sometimes the sides) of their containers with this number prior to shipment (Figure 7); this allows quick and convenient identification of the material upon arrival at the facility. Once a numerical code has been assigned to a waste stream, the generator can then use it as a reference to obtain cost information, access approval status, schedule shipments, or trace waste management activities and ultimate disposition.

In at least one state, South Carolina, a waste specific Authorization Number is assigned by the TSDF to each unique hazardous waste that will ultimately be treated or disposed of in that state; see Table 2 for sample code. This number is submitted on an Authorization Request Form (Figure 2) along with a description of the waste and other information (e.g., EPA/ State waste codes, DOT hazard class, and physical/chemical data) to the South Carolina Department of Health and Environmental Control, Bureau of Solid and Hazardous Waste. In Figure 2, the Authorization Number consists of a two letter TSDF prefix, followed by, for example, a generator code and a specific waste stream code (Note: the State of South Carolina

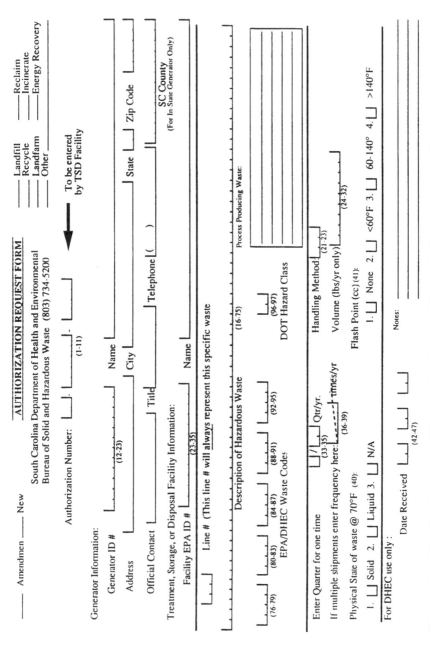

Figure 2. South Carolina Department of Health and Environmental Control Authorization Request Form.

allows the TSDF to use its own internal coding system for the assignment of authorization numbers). The Authorization Number helps the S.C. Department of Health and Environmental Control track solid and hazardous waste shipments to and within their state.

Generator Systems

Generator waste identification systems are rather similar to those used by commercial TSDFs to track incoming shipments. In practice, the first step is to adopt a facility-wide system that codifies wastes by point of generation. Each waste is assigned a unique numeric or alphanumeric code often referred to as the Stream Number or Waste Identification Number (Waste ID No.). Once assigned, the Waste ID No. plays a key role in on site, as well as off site, waste material tracking and cost accounting.

GENERATOR WASTE ID NUMBERS: FUNCTION AND USE

Why should generators use Waste Identification Numbers? The development and use of waste identification numbers represents the first step toward organizing data collection efforts. The waste identification number offers a convenient way to trace all management activities and costs directly back to the precise point of generation. From a recordkeeping perspective, Waste ID Nos. may be viewed as a sort of "shorthand" and, in some ways, are comparable in use to "scientific notation." Scientific measurements and calculations, for example, can involve very large numbers. To express these numbers more conveniently, scientific notation is often used. Similarly, codes are often used in place of lengthy process titles, waste descriptions or names of generating departments. Rather than record "Cutting oil/floor absorbent from small Milwaukee milling machine No. 10, second floor Milling Department" over and over again, it would be more convenient to refer to a standard, abbreviated source identifier.

Employed as an Accounting Tool

As an accounting tool, Waste ID Nos. are employed as tracers in the waste management cycle to track the movement of each unit of waste from the point of generation through transportation, treatment and disposal. By assigning a unique identifier to each waste, unit tracking can be efficiently accomplished and "fully-loaded" waste management costs assigned directly to the product, process, operation or department of origin. Where similar wastes are generated at separate locations within the same department or if their departments of origin are different, the Waste ID No. provides a means to make this distinction. For an illustration of this point, refer to the Case Study presented on the following page.

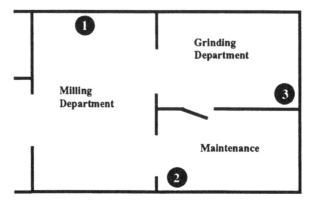

Figure 3. Satellite accumulation of three similar types of oily wastes.

Avoiding Common Problems

The use of source identifiers can help managers avoid in-house problems associated with "orphan" (of unknown origin) drums or containers. Instances, for example, where the department of origin (or point of generation) is not easily determined can be avoided by adopting a routine procedure for marking containers with the Waste ID No. at the time accumulation begins. This marking procedure can save on costly analytical fees associated with the characterization of "unknowns" and will also allow containers, found to be off-spec or improperly packaged by some distant off site TSDF, to be quickly and conveniently traced directly back to the responsible party or department of origin.

Waste ID Nos. are also very useful as markers in contaminant tracing, for tracking waste minimization progress, indexing data and as a basic accounting tool in the performance of chemical mass balance calculations — such as those performed pursuant to the Toxic Chemical Release Reporting provisions of the Emergency Planning and Community Right-to-Know Act of 1986. When used properly, waste identification numbers will play a key role in the success of your in-house tracking and cost accounting program.

Case Study: Hypothetical

In the facility shown in Figure 3, there are three satellite accumulation* areas. A 55-gallon open-head steel drum is stationed at each satellite lo-

---

* Satellite accumulation is the term often used to refer to the EPA regulation that states: "A generator may accumulate as much as 55 gallons of hazardous waste or one quart of acutely hazardous waste listed in §261.33(e) in containers at or near any point of generation where wastes initially accumulate, which is under the control of the operator of the process generating the waste without a permit or interim status ..." [6]

cation for the collection of floor absorbent contaminated with machine oil. On an average, the Milling Department generates eight (8) drums of oily waste every 90 days, the Grinding Department four (4) drums and the Maintenance Department two (2) drums. At the time of shipment, the generator adds all oily waste drums together and records the total or "aggregate" quantity as one entry on the shipping paper; e.g., "Oil-Contaminated Floor Absorbent, 14 Drums." In this case, containers are not marked by point of generation and no attempt is made to track the number of drums generated by each department over time. Since efforts have not been made to track wastes by department of origin, transportation and disposal costs cannot be equitably assigned. Since costs are not charged back to the generating department based on the rate of generation, there is little direct economic incentive for departments to reduce floor spillage. Also, if a drum is later discovered to have been cross contaminated with an industrial solvent, for example, there is no system in place to assign "departmental ownership" or to assist in the investigation of the incident.

DuPont Example

In their work "Implementing Waste Minimization Programs in Industry,"[7] Hollod (Conoco Inc.) and Beck (E.I. DuPont) describe the value of establishing a waste material tracking system with respect to identifying opportunities and measuring progress in waste minimization:

> A tracking system can be used to identify waste reduction opportunities and provide essential feedback to successfully guide future efforts. Considerations to volume of waste, cost of handling, regulatory impact, product life cycle, marketing opportunities, and basic manufacturing processes can be factored into an algorithmic function and coupled with a tracking system to identify opportunities. The identification process would allow a business team to maximize the technical and capital resources available and direct them to the part of the business that needs them the most to improve its overall performance.

> ...The tracking function or recordkeeping at a minimum should record and *identify the generator or "owner" of the waste* [emphasis added], associated volumes, cost of managing the waste, and the waste reduction method being used to reduce that particular wastestream (Hollod and Beck, "Implementing Waste Minimization Programs in Industry." Reprinted with permission of McGraw-Hill, Inc. © 1990).

A table included in their article shows that DuPont used five (5) digit Waste ID Nos. to key data back to "production areas" that generated waste.

For example, in a spreadsheet format, the company tracked (1) the quantity, 50,000 pounds per year, of an "acidic, spent catalyst," (2) associated management/disposal costs, at $42,000 per year, and (3) the method for minimizing this waste, by department of origin. Chapters 5 and 6 describe a framework for tracking these and other variables by point of generation.

SYSTEM DEVELOPMENT

In practice, the overall value of a structured waste identification system is directly related to the number of waste streams generated. That is, the greater the number of waste streams, the more valuable Waste ID Numbers become; the converse of this, however, is also true.

Key Word Identifiers

In companies where relatively few waste streams are generated, the source identifier may not be a number at all, but rather a short title or key word that refers to the type of waste or point of origin. If, for example, a company generates petroleum-based lubricant from a single source and spent mineral spirits from one cold cleaning operation, source identifiers could simply be recorded as "Used Oil" and "Spent Solvent." In this case, the company generates only two types of waste and the identifier is the name commonly used to refer to the material.

If your company is a small generator with only a few types of wastes, what you decide to call the waste is not as important as how you use it. Your waste identifier, regardless of how simple it is, should be used consistently throughout your entire tracking and cost accounting program.

Numeric Waste Identifiers

In large companies where many different types of waste are generated or where similar wastes are generated at different locations, a more structured approach to identifying wastes by point of generation is needed. One of the more common types of source identification system relies on the use of serially increasing alphanumeric designations.

Alphanumeric/Numeric Codes
Alphanumeric source identifiers are those which employ both alphabetical and numeric symbols. A very simple alphanumeric waste identification system can be constructed as follows. If, for example, you want to track your wastes by department of origin and your maintenance department generates three types of waste (namely, waste oil, spent trichloroethylene, and paint waste), then serially increasing alphanumeric Waste ID

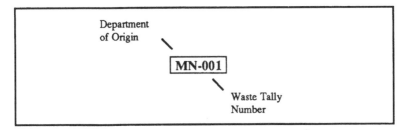

Figure 4. An example of a simple alphanumeric waste identification number.

Nos. could be assigned as MAIN-01 or MN-01, where MAIN/MN = Maintenance and 01 = waste oil, MN-02 for spent trichloroethylene and MN-03 for paint waste. The maintenance department in your organization may also, however, be known or commonly referred to as Department 14. Waste ID Nos. in this case could also be recorded numerically as 14-01, 14-02 and 14-03; for departments which generate many types of wastes an additional place holding zero (0) could be added, for example 14-001.

In Figure 4 and as with all source identifiers, the base unit of the Waste ID No. is that component which identifies the department or operation of origin. The waste tally number, on the other hand, provides a means to distinguish among wastes generated from the same department or operation. [Note: Additional thoughts on this subject are provided in *Tracking Additional Information,* Chapter 6.]

Applications in Larger Organizations

Now, the source identifiers discussed so far are rather simple and will work well for most generators. Some of the larger, more complex facilities, however, may require the assignment of waste identification numbers which contain additional digits. In its publication entitled *Waste Minimization: Manufacturers' Strategies for Success,* the National Association of Manufacturers (NAM)[8] presents examples of a ten-digit, hyphenated alphanumeric "Stream ID Number." The company reviewed in the NAM publication assigned the number A-123-1-682-03 to a spray booth cleaning operation waste — xylene contaminated rags and filters (4.5% xylene) — that was generated at a rate of about 2 drums per month. Here, the stream identification number "represents the plant area, building, floor, department and waste tally number."

Final Considerations

In the final analysis, the method you use to identify your waste should be simple, easy to communicate and reflective of your company's organizational structure. Remember, the Waste ID No. is the means by which information will be accessed and unit tracking accomplished across each step in your accounting program, as shown in Figure 5.

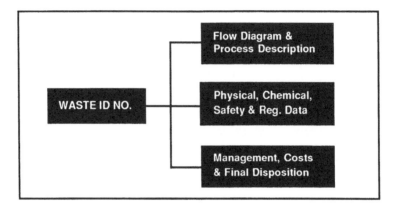

Figure 5. Organization and retrieval of information by waste ID number.

CREATING AN INVENTORY

Once you have formulated an appropriate identification system, the next task is to identify and define the universe of all waste materials generated at your facility. This involves performing a facility-wide audit to determine each process, operation and activity that generates solid and hazardous waste. The product of this audit should be a written inventory organized and tabulated by point of generation. All wastes that are containerized or shipped in bulk to an off site treatment, storage or disposal facility should be included in the inventory. In addition to hazardous wastes, solid wastes such as asbestos, demolition debris, non-hazardous manufacturing by-products and residues should also be listed.

RECORD REVIEW AND DEPARTMENT INTERVIEWS

In order to develop a complete inventory, it may be helpful to review historical records. Existing information including hazardous waste manifests, generator biennial reports, bills of lading, spill reports, accounts payable records, commercial TSDF contracts/waste stream summaries and other internal documents will be useful in this effort. It may also be helpful to interview department representatives such as area foremen and operators. These personnel are most familiar with the waste generating processes (especially the frequency and occurrence of special or "one-time" wastes such as those resulting from the cleanup of chemical spills) and have first-hand knowledge of facility operations.

DEVELOPING A WRITTEN INVENTORY

Once a complete inventory of all known wastes has been taken, the next step is to develop a comprehensive written list. Each type of waste is assigned a unique waste identification number and recorded categorically, by department of origin, for example, on an index sheet; see Figure 6. For companies that generate many types of waste, the information may be stored electronically using a simple computer spreadsheet program.

In Figure 6, wastes are listed by department of origin with an entire work sheet dedicated to the Maintenance Department. In practice, wastes originating from more than one department may be combined onto a single record. In either case, be sure to leave enough space for (1) new wastes not currently generated or anticipated (for example, wastes generated from new product development, accidental spills or obsolete inventory), and (2) additional wastes that may be identified in Chapter 3 as the result of a facility walkthrough or process review.

DOCUMENTATION

Recording* data in the form of a solid and hazardous waste index is rather simple. Waste identifiers are recorded in a Source/ID column. Information of special interest, vendor approval codes for example, can be entered into a second column; see Figure 6.

In the example provided in Figure 6, a third column is entitled DE-SCRIPTION OF WASTE. Here, a short descriptive title is assigned to the waste generating process, operation or activity. In practice, enter a common name that will clearly identify the waste for future reference. It is important to assign descriptors that are concise yet "telling" and adequately detailed. Descriptors should also be consistent with the ones used on vendor profile sheets, government or company surveys and reports. This practice will provide continuity throughout your tracking and cost accounting efforts. Together, the Waste ID Number and descriptor will provide positive identification of the point of generation.

The reverse side of the index sheet (not shown in Figure 6), or fourth column in the case of electronic systems, can be used to expand upon "waste descriptions" or record itemized lists of materials. Short narrative descrip-

---

* Note: For manual recordkeeping systems, custom tracking and cost accounting work sheets are easily developed. The work sheets presented in this book (Figures 6, 12, 13 and 30 through 34, for example) were designed using an Aldus Pagemaker® computer graphics software program. Once designed, work sheets and recordkeeping forms may be produced using a laser printer, capable of at least 300 dpi, and then copied (or commercially printed) on 70 lb. paper stock with holes drilled in the margins for storage in a three-ring binder. See also discussion on "Electronic Systems" Chapter 6.

**Solid & Hazardous Waste
Index**

SOURCE/ID NO.	APPROVAL CODE(S)	DESCRIPTION OF WASTE
MAINTENANCE		
MN - 001	U26261, L-4526	Waste lubricating oil from fork truck maintenance
MN - 002	◄	Spent 1,1,1- trichloroethane from parts cleaning
MN - 003	A Second column can be used to record Vendor Approval Codes, Product Codes, Regulatory Status, Volume Totals or other information.	Paint waste: obsolete inventory * See reverse
MN - 006		Floor trench clean-out solids/South sump
MN - 007		Friable Asbestos from pipe jacket; repair work shop boiler
MN - 008	ST-00009-9542	Oil spill clean-up debris: 40% No. 6 Fuel Oil, Sand, Rags
MN - 009	BH4898	Forktruck Batteries * See reverse for itemized list

Figure 6. Sample of an index sheet dedicated to a single department of origin.

tions that summarize how non-process wastes — such as those that result from spills, special operations or maintenance activities (e.g., wastes identified as MN-001, MN-007 and MN-008 in Figure 6) — are first generated should also be recorded here; this will ensure that the history of origin for each waste is not lost. Information and data on the origin of process wastes will be documented as outlined in Chapter 3.

TRACKING BY ID NUMBER

CONTAINER MARKING

General Guidance

Now that your in-plant system has been established, the Waste ID No. should be routinely marked on each container of waste — either on a label with indelible ink or directly on the container itself with a stenciling system, grease pencil or some other implement.* Be certain that this writing does not interfere with state and federal marking or labeling requirements. The United States Department of Transportation, for example, requires hazardous material markings to be "displayed on backgrounds of sharply contrasting colors, unobscured by labels or attachments, and located away from other markings, such as advertisements which could reduce their effectiveness" (49 CFR § 172.304, noted in Table 3).

Federal Marking and Labeling Requirements

Table 3 outlines the basic federal requirements for container marking and labeling. In order for a generator to be in compliance with federal law, each container of hazardous waste must be labeled and marked in accordance with both the U.S. EPA and U.S. DOT regulations. Where a container is improperly labeled or marked, the generator may be held liable for penalties assessed by either or both agencies.

---

* Drum marking materials and labeling systems are commercially available. Preprinted DOT labels and EPA markings, for example, can be purchased for most applications. For unusually large applications, stand-alone electronic systems complete with computer, software and printer can be purchased for printing labels inhouse. Sources such as the Thomas Regional Directory (published by Thomas Publishing Company, Inc., One Penn Plaza, New York, NY 10001 — circulation is free to qualified individuals who are involved in purchasing, specifying, or recommending, and whose company is located within the region from which the request is made), buyer's guides on environmental products (such as Pollution Equipment News' *Buyers Guide* published by Rimbach Publishing Inc. of Pittsburgh, PA) and trade publications on product identification systems for industry are good places to begin your search for drum labeling and marking materials and systems.

Table 3
Summary of Federal Label and Marking Requirements for Containers of Hazardous Waste

U.S. ENVIRONMENTAL PROTECTION AGENCY
Labeling
1. Before transporting or offering hazardous waste for transportation off site, a generator must label each package in accordance with the applicable DOT regulations on hazardous materials under 49 CFR Part 172 (DOT labeling requirements listed below).

Marking
1. Before transporting or offering hazardous waste for transportation off site, a generator must mark each package of hazardous waste in accordance with the applicable DOT regulations on hazardous materials under 49 CFR Part 172 (DOT marking requirements listed below).
2. Before transporting or offering hazardous waste for transportation off site, a generator must mark each container of 110 gallons or less used in such transportation with the following words and information displayed in accordance with the requirements of 49 CFR 172.304:

HAZARDOUS WASTE — Federal Law Prohibits Improper Disposal. If found, contact the nearest police or public safety authority or the U.S. Environmental Protection Agency.
Generator's Name and Address -----------------.
Manifest Document Number -----------------.
3. The date upon which each period of accumulation begins is clearly marked and visible for inspection on each container.
4. While being accumulated on site, each container is labeled or marked with the words, "Hazardous Waste."

U.S. DEPARTMENT OF TRANSPORTATION
Labeling
1. 49 CFR § 172.400 General labeling Requirements — specifies which packages and containment devices must be labeled.
2. 49 CFR § 172.400a Exceptions from labeling — specifies instances when labels are not required.
3. 49 CFR § 172.401 Prohibited labeling — specifies conditions for use and prohibitions on use.
4. 49 CFR § 172.402 Additional labeling requirements — lists additional requirements for subsidiary hazard labels, display of hazard class on labels, cargo aircraft only label, radioactive materials and class I (explosive) materials.
5. 49 CFR § 172.403 Radioactive material — specifies labeling requirements for radioactive materials.
6. 49 CFR § 172.404 and § 172.405 provide labeling requirements for mixed and consolidated packaging, and for authorized label modifications.
7. 49 CFR § 172.406 Placement of labels — provides general requirements and specifies placement of label, contrast with background, duplicate labeling and visibility.

Continued

Table 3 (Continued)

U.S. DEPARTMENT OF TRANSPORTATION
  Labeling
    8. Notable rules are that labels must be :
      a. Printed on or affixed to a surface (other than the bottom) of the package or containment device containing the hazardous waste
      b. Located on the same surface of the package and near the proper shipping name marking
      c. Clearly visible and may not be obscured by markings or attachments
    9. 49 CFR § 172.407 through § 172.450 contain specifications for each authorized label (e.g., oxidizer, poison, organic peroxide).

  Marking
    1. 49 CFR § 172.300 Applicability — general applicability standards.
    2. 49 CFR § 172.301 General marking requirements for non-bulk packagings — includes requirements for proper shipping name and identification numbers, exemption packaging, consignee/consignor name and address, previously marked packagings and marking exemptions.
    3. 49 CFR § 172.302 General marking requirements for bulk packagings.
    4. 49 CFR § 172.303 Prohibited marking rules.
    5. 49 CFR § 172.304 Marking requirements — specifies that all required markings be:
      a. Durable and in English and printed on or affixed to the surface of a package or on a label, tag, or sign;
      b. Displayed on a background of sharply contrasting color;
      c. Unobscured by labels or attachments; and
      d. Located away from any other marking (such as advertising) that could substantially reduce its effectiveness.
    6. 49 CFR § 172.306 Removed and reserved.
    7. 49 CFR § 172.308 Authorized abbreviations.
    8. 49 CFR § 172.310 through § 172.338 contain specifications for selected hazardous materials (e.g., radioactive, poisonous) and special provisions for cargo/ portable tanks, identification numbers, marine pollutants and other unique marking requirements.

Source: Titles 40 Part 262 and 49 Part 172 of the Code of Federal Regulations.

The EPA essentially requires that containers of 110 gallons or less be marked with the accumulation start date, the words "Hazardous Waste" while being accumulated on site, the generator's name and address, the manifest document number and a specific notice intended to thwart the improper disposal of hazardous waste (see Table 3, U.S. EPA marking requirements). The U.S. DOT requires generators to comply fully with its marking and labeling standards, since all federally regulated hazardous wastes are automatically regulated by the U.S. DOT as hazardous material.

Marking and Labeling Practice

As discussed earlier, in order to properly track hazardous waste each unique waste identification number must be clearly marked on the container. Additionally, TSDFs often require generators to mark the top of each container with a unique vendor waste code or profile number. It is important that these numbers do not interfere with, detract from or in any way obscure federal labeling and marking requirements.

All marking and labeling requirements, whether or not required by government agencies, can be grouped into one or more of the following categories (see Figure 7): (1) TSDF marking requirements (if specified by vendor), (2) U.S. Environmental Protection Agency marking requirements, (3) U.S. Department of Transportation marking and labeling standards, (4) state regulatory marking standards (where existent), and (5) in-house waste tracking marking(s).

Of these five categories, two may not be required — TSDF and state regulatory marking requirements; your state environmental protection agency may or may not have promulgated container marking regulations that are in addition to the federal requirements. Your in-house waste tracking marking, though not required by an outside agency or organization, should be used routinely and become a permanent part of your waste management program.

A simple labeling and marking layout is presented in Figure 7 to help guide you in the preparation of your containers for shipment. Additionally, Lion Technology Inc., a leader in hazardous waste compliance training, has developed a rather innovative marking system that is designed to "implement DOT and EPA marking requirements for the shipment of hazardous waste in drums while minimizing the shipper's liabilities." The Lion Drum Marking System incorporates a waste tracking component and is simple to use. Additional information can be obtained from Lion Technology Inc., P.O. Drawer 700, Lafayette, New Jersey 07848.

CONTAINER STORAGE

In general, there are no specific federal regulations* that require generators to store hazardous waste containers indoors although, for waste tracking purposes, this is the most prudent management method. Contain-

---

* Note: (1) Be sure to check state and local regulations for hazardous waste storage requirements. State hazardous waste enforcement agencies are permitted, as authorized by federal law, to adopt more stringent waste management standards. (2) Generators that store hazardous waste for more than 90 days are required to comply with the more comprehensive containment design standards of 40 CFR § 264.175. Also, generators who choose to store containers inside a "containment building" must comply with the design and operating standards of 40 CFR 265 Subpart DD.

23

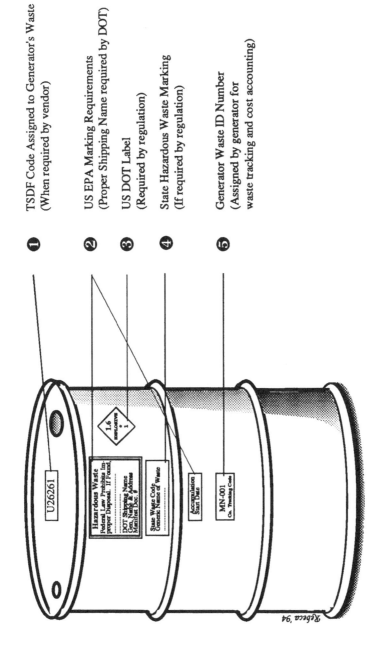

① TSDF Code Assigned to Generator's Waste
(When required by vendor)

② US EPA Marking Requirements
(Proper Shipping Name required by DOT)

③ US DOT Label
(Required by regulation)

④ State Hazardous Waste Marking
(If required by regulation)

⑤ Generator Waste ID Number
(Assigned by generator for
waste tracking and cost accounting)

Figure 7. Sample labeling and marking schematic for waste tracking and regulatory compliance.

ers that are stored outdoors are continuously exposed to the elements. Markings, including waste identification numbers, and labels may not be durable enough to withstand extreme weather conditions. When markings that identify a container's place of origin are no longer legible, an "orphan" drum may result, causing the environmental manager to either (1) expend additional administrative time in trying to locate the generator, or (2) organize a sample collection and analysis effort to identify, through (often costly) laboratory testing, waste material composition.

In the book *Prudent Practices for Disposal of Chemicals from Laboratories*,[9] the National Research Council offers sound advice on storing and protecting containers of hazardous waste from the elements:

> Waste containers should always be stored in an area or facility where the containers are protected from adverse weather. Extremes in heat can cause pressure buildup in containers with volatile liquid contents. Alternate heating and cooling may cause containers to breathe if they are not tightly sealed. If water is left standing on the container, it can be drawn into a container during a cooling cycle, resulting in container corrosion or internal chemical reactions. Adverse weather conditions can also cause deterioration of labels, tags, or other markings, which can create a hazard and could necessitate reanalysis of container contents.
>
> A properly designed storage facility should have a roof to protect waste from sun and precipitation. Warning signs should be posted and walls or fences erected to protect against unknowing or unauthorized entry. The facility must, however, have proper ventilation and must provide access for fire or other emergency equipment. The facility should be located away from areas of high work density but close enough to be useful and to allow proper surveillance and security (Reprinted with permission from PRUDENT PRACTICES FOR DISPOSAL OF CHEMICALS FROM LABORATORIES. Copyright 1983 by the National Academy of Sciences. Courtesy of the National Academy Press, Washington, D.C.).

Though there are no specific requirements calling for indoor storage, EPA regulations do address storage with respect to the condition of containers, the compatibility of wastes with the container, management of containers, inspections, and general and specific requirements for ignitable, reactive, and incompatible wastes. These storage requirements are listed and summarized in Appendix A. EPA has also published guidelines for product and raw material storage. Those guidelines that are equally applicable to the storage of containerized hazardous waste are presented in Appendix B.

As a general rule, it is easier to maintain compliance with federal marking, labeling and management requirements when containers are stored indoors or are otherwise protected from the elements. [Note: Markings and labels that are subject to the elements, even for 90 days, can fade in appearance. If the loss in appearance is too dramatic not only do you risk being unable to identify the contents of the drum, but you may also be in violation of the law. EPA regulations require that the accumulation start date is "clearly marked and visible for inspection" and that each container is "marked clearly" with the words "Hazardous Waste." In addition, DOT regulations require that labels are "clearly visible" and that markings are "durable" and "displayed on backgrounds of sharply contrasting color" (see Table 3 for more detail)].

WASTE STREAM SEGREGATION AND LOGS

The key to waste material tracking by point of generation is waste stream segregation. It is very important that wastes, where practicable, are collected and stored separately by generating source. Waste stream segregation not only facilitates material tracking/cost accounting efforts, but also (1) increases the value of wastes intended for on/off site recycling or reclamation, (2) affords greater opportunity for the implementation of pollution prevention measures, and (3) may result in lower treatment or disposal costs.

In the case of waste mixtures (i.e., the intentional commingling of wastes in a single container or tank) a running log should be kept on the volume and origin of each waste entering the storage container/tank; a discussion of tracking the components of waste mixtures and the use of running logs is provided in Chapter 4. At the time of each shipment, a tally of the total amount of each unique waste by generating source is made. As outlined in Chapters 5 and 6, results are then recorded in a solid and hazardous waste shipment journal for that day.

As a final note, when recordkeeping responsibilities are transferred to a new individual, every effort should be made to maintain consistency in documentary style and content. Also, whenever a substantive change is made it should be documented across each step in your waste tracking and cost accounting program.

REFERENCES

1. **40 Code of Federal Regulations** §260.10.
2. **ibid.**
3. **ibid.**
4. **Egan, P.,** provided assistance in compiling Table 2. Mr. Egan is associated with Clean Harbors Environmental Services Co., 1200 Crown Colony Dr., Quincy, MA 02269-9137.
5. **Locklear, J.L.,** personal communication (as written correspondence), Laidlaw Environmental Services of South Carolina, Inc., Route 1, Box 255, Pinewood, SC 29125. August 19, 1994.
6. **40 Code of Federal Regulations** §262.34 (c).
7. **Hollod, G.J., and Beck, W.B.,** Implementing Waste Minimization Programs in Industry, in H.M. Freeman, ed., *Hazardous Waste Minimization,* McGraw-Hill Publishing Company, Inc., 1221 Avenue of the Americas, New York, New York 10020. 1990.
8. **National Association of Manufacturers,** *Waste Minimization: Manufacturers' Strategies for Success,* 1331 Pennsylvania Ave., NW, Washington, D.C. 1989.
9. **National Research Council,** *Prudent Practices for Disposal of Chemicals from Laboratories,* National Academy Press, 2101 Constitution Ave., NW, Washington, D.C. 20418. 1983.

Chapter 3

PROCESS, OPERATION AND MATERIAL REVIEWS
Analysis of Waste Generating Processes and Documentation

INTRODUCTION

Chapter 2 focused on identifying and defining the universe of waste materials generated at your facility. Now it's time to do some detailed investigative work. In this next step, a foundation for your waste tracking and cost accounting program will be built by documenting precisely *how* and *where* each waste is generated. The overall approach is rather simple. First, a thorough analysis of each process, operation and activity that generates solid and hazardous waste is performed. Second, the results of each analysis, with special reference to materials used, released and generated as waste, are carefully documented in the form of (1) process flow/block diagrams, and (2) written process descriptions. Once recorded, the information and data collected in this step will then serve as a basis and valuable resource for future tracking, cost accounting and material evaluation efforts.

THE IMPORTANCE OF UNDERSTANDING YOUR
WASTE GENERATING PROCESSES

The knowledge gained during process reviews can be used for a variety of purposes, many of which are discussed in the sections and chapters which follow. Chief among the potential uses of such information and data is the characterization of wastes. Classifying wastes in accordance with federal law[1] represents one type of material evaluation effort that is very much dependent upon the specifics regarding "how" wastes are first generated. If a waste is improperly characterized, for example, all subsequent tracking and accounting steps will carry this error forward — possibly resulting in a financial penalty. The first step in any characterization effort is to understand the waste generating process or operation. This knowledge is then applied to hazardous waste listings and characteristic criteria.

As a case in point: the precise RCRA classification of certain "waste" solvents* is entirely dependent upon whether the material was first used as

---

* Spent solvents listed as hazardous waste in 40 CFR 261.31 include only those solvents that are used for their "solvent" properties (i.e., to solubilize or mobilize other constituents). The current listings do not encompass process wastes where solvent constituents are used as reactants or ingredients in the formulation of commercial chemical products. [2,3]

27

Table 4
RCRA Classifications for Hazardous Waste Containing TCE [a]

Hazardous Waste No.	Primary Constituent	When Subject to Classification as Hazardous Waste
F001	Trichloroethylene	Spent, when used in degreasing.
F002	Trichloroethylene	Spent, when used as a solvent in operations other than degreasing.
U228	Trichloroethylene	When (1) discarded as a commercial chemical product, (2) discarded as an off-specification commercial chemical product, (3) remaining in a container as residue or in an inner liner removed from a container, and (4) when present in any residue or contaminated soil resulting from the cleanup of a spill into or on any land or water. [b, c]
D040	Trichloroethylene	When not listed as an F001, F002 or U228 hazardous waste and exhibiting the Toxicity Characteristic for TCE (regulatory level equal to or greater than 0.5 mg/l).

[a] Source: Title 40 Code of Federal Regulations (CFR) Part 261.

[b] See 40 CFR § 261.33 for complete text.

[c] The phrase "commercial chemical product" refers to a chemical substance which is "manufactured or formulated for commercial or manufacturing use which consists of the commercially pure grade of the chemical, any technical grades of the chemical that are produced or marketed, and all formulations in which the chemical is the sole active ingredient. It does not refer to a material, such as a manufacturing process waste [that contains trichloroethylene]. When a manufacturing process waste is deemed to be a hazardous waste because it contains [trichloroethylene] it will be [otherwise] listed" (Source: 40 CFR § 261.33).

a cleaning agent, ingredient, diluent, reactant, or as a medium for a chemical reaction. Depending upon how a solvent is first used, it is entirely possible for chemically similar wastes to be classified very differently under EPA regulations. This difference in material classification can not only affect the focus of a company's waste management initiative, but can also mean the difference between whether a company is operating in compliance or noncompliance with federal law.

The trichloroethylene (TCE) and chlorofluorocarbon (CFC) examples that follow underscore the importance of understanding your waste generating processes and operations in terms of how chemicals are used and how wastes are generated. Though these examples are rather straightforward, other listed solvents can often be more difficult to classify, especially those used in chemical manufacturing where the precise chemistry of the process must be understood in order to determine whether the solvent is used as a diluent, reactant or medium for a chemical reaction.

Since the technical reviews performed in this chapter directly affect waste material classification, tracking and other elements of your hazardous waste management program, you should set aside sufficient time to review and confirm the accuracy of all data to be collected.

Trichloroethylene Example

Table 4 shows that trichloroethylene (when spent or discarded) may be classified and regulated as a hazardous waste according to four different scenarios, each classification being dependent upon how the chemical is first used or how the trichloroethylene-bearing waste is first generated. If for example, TCE is used as a degreasing agent, as in Figure 9, and is "spent", or can no longer be used for its intended purpose, then it would be regulated as an F001 hazardous waste. If the same virgin material is used for its solvent properties (to dissolve another material other than in a degreasing operation), however, and is then discarded, it would at that time be assigned the Hazardous Waste No. F002. Should the TCE be discarded as an unused commercial chemical product or before use as a container or spill residue, it would be classified as a U228 hazardous waste. Finally, when all of the other listings do not apply and when a TCE-bearing waste exhibits the Toxicity Characteristic for TCE, it is then regulated as a D040 hazardous waste.

Chlorofluorocarbon Example

A second example can be found in the 28 July 1989 Federal Register (54 FR 31335). In this notice, EPA discusses how to classify chlorofluorocarbons based on their use — with special emphasis on the applicability of RCRA hazardous waste regulations to CFC refrigerants. The core rationale used by EPA in their discussion on waste characterization is reprinted below:

> The applicability of RCRA [hazardous waste] regulations to CFCs is limited to three basic scenarios: (1) Where CFCs are used as solvents and the wastes containing the CFCs meet the F001 and F002 listing descriptions, (2) where either dichlorodifluoromethane (CFC-12) or trichloromonofluoromethane (CFC-11) is an unused commercial chemical product, off-specification commercial chemical product, inner liner or container residue, or spill residue that is (or is intended to be) discarded, or (3) where CFCs are solid wastes that exhibit a characteristic of hazardous waste.
>
> First, the spent solvent listings found at 40 CFR 261.31 (specifically, CFCs listed under F001 and F002) apply solely to wastes

containing listed solvents when they are used for their solvent properties. CFCs used as refrigerants are not typically subject to the spent solvent listings because, as refrigerants, the CFCs are not used as solvents. Second, the U-listings found at 40 CFR 261.33(f) apply to commercially pure grades, and formulations in which the listed chemical is the sole active ingredient, and do not apply to chemicals that have been used for their intended purpose. Thus, CFC refrigerants that are removed from a refrigeration system and are reclaimed would not be classified as "commercial products," but rather would be classified as "spent materials." If the CFC refrigerants were not used for their solvent properties, they could not be F001 or F002 wastes, and thus, these spent materials could only be hazardous wastes under the characteristics of 40 CFR 261.21-261.24.

As a spent material, a CFC refrigerant is a solid waste. It is therefore the generator's responsibility to test the waste or apply knowledge of the waste to determine whether the waste exhibits a characteristic of a hazardous waste. The characteristics of a hazardous waste [i.e., ignitability, corrosivity, reactivity, or EP Toxicity (Author's Note: now the Toxicity Characteristic)] are found at 40 CFR 261.21-261.24.

The Agency has previously determined that CFC refrigerants are not likely to exhibit a characteristic of a hazardous waste; however, the Agency maintained reservations regarding the characteristic of corrosivity. EPA was concerned about the possible formation of hydrochloric acid due to the breakdown of the CFCs at high compressor temperatures. EPA has since received data demonstrating that the conditions under which CFC refrigerants would break down and form hydrochloric acid, while theoretically possible, are not a practical possibility during normal use.

COLLECTING PROCESS DATA

In the sections which follow, you will be asked to conduct a basic review of each process, operation and activity that results in the generation of waste. Specific guidance is given on data collection and documentation. Data collection efforts will focus on the details concerning the origin of each waste. Technical data relating to feedstock materials, process and operational methods, chemical release and waste production will be examined and systematically recorded by point of generation.

Records should be kept during all data collection and point of generation review activities; a sample process flow sheet is shown and its use described in the Detailed Guidance section of this chapter. Keeping clear and concise records will not only assist you in subsequent tracking activities, but will also help to demonstrate compliance with federal law by serving as a basis for your waste characterization efforts when laboratory testing is not required. Also, when waste generation data are organized and adequately detailed, precise unit tracking and fully-loaded waste management cost accounting will proceed more efficiently.

SMALL TO MEDIUM-SIZED FACILITIES

In small to medium-sized facilities, data collection efforts may be carried out by one or two individuals. As part of the data collection process, investigators should visit the site of generation to observe waste generating processes and material handling techniques. Where possible, "the visit should coincide with [the process or operation generating the waste] such as equipment cleaning. It can be useful to observe operations in more than one shift, particularly if activities differ from one shift to another or if waste generation is highly dependent on employee activities. A visit also allows [the investigator] to discuss current housekeeping practices with operational staff and to clarify ambiguous diagrams and plans." [4]

The walkthrough (or site visit) provides the investigator with an opportunity to acquire first-hand knowledge of the processes and operations that generate waste. New information may be obtained and existing data checked for accuracy and completeness. In his guide, *Profiting from Waste Reduction in Your Small Business*, Wigglesworth[5] discusses the value of a walkthrough in the context of identifying opportunities to reduce waste:

Investigators "may think that they are familiar enough with the facility to eliminate this process [the walk-through]. But familiarity can be as much a hindrance as a help. A walkthrough gives the [investigator] a chance to objectively look at the facility. The basic observations made during the walkthrough will provide invaluable information on waste production and reduction. It is important to record your visual observations and discussions held during the walkthrough. Drawing a simple diagram will sometimes provide you with ideas on how to modify your business operations to reduce waste."

LARGE FACILITIES

In larger facilities, process reviews may be more efficiently accomplished through a coordinated interdepartmental effort. An interdepartmental task force or project team may be assembled or process flow sheets (refer to Detailed Guidance section of this chapter) distributed to a designee

Table 5

Resources and Methods for Obtaining Waste Generation Data

Data Collection Sheets

These are forms designed to obtain specific data and are typically completed by those who are closest to the process or operation. In order to determine, for example, the total amount of spent TCA processed or still bottoms generated per unit time (Figure 9), an operator may be requested to keep a daily log on a specially designed data collection work sheet.

Information from Specialists

Generally speaking, the person who is or has been closest to the waste generating process or operation is the expert. This person or persons should be consulted throughout your data collection efforts. Specialists can help you develop your understanding of the waste generating process and provide valuable insight into operational methods or potential pollution prevention techniques.

Facility Records/Historical Data

Information often exists in the form of operating manuals, diagrams (e.g., block, flow or piping and instrumentation), process sheets and company reports. Data collected in the past are valuable and can serve as a good reference point from which to begin your investigative work. Be careful, however, to check the accuracy of all historical data against current operating practices and information.

Instrumental Analysis

New data may be generated by subjecting representative samples of raw materials, materials in process or waste materials to a qualified laboratory for analysis. Guidance on waste material sampling and analysis is given in Chapter 4.

in each waste generating department. With some basic guidance, task force members or departmental designees would be able to conduct a walk-through and complete a flow sheet for each waste generating process, operation, and activity. Completed forms could then be returned to a central office or program coordinator for final review and approval. In this case, final approval would be contingent upon a quality check in which the program coordinator visits the site of generation to confirm all flow sheet data. Once the forms have been collected and approved, they are organized by point of generation in the order in which they appear in the Solid & Hazardous Waste Index, Chapter 2. Key process and operational information may then be quickly and conveniently accessed by waste identification number.

WASTE GENERATION DATA

As you proceed with your data collection efforts, and throughout the development of your entire tracking and cost accounting program, there will be numerous times when process specific waste generation data and information are needed. The resources and methods available to you for obtaining this data generally fall into one of the four categories listed and summarized in Table 5. Keep in mind, however, that your goal throughout all data collection efforts should not be to simply gather more data, but to collect technically sound and useful data.

Once data have been collected, they must then be carefully organized and clearly recorded. General guidance is given in the following section on how to organize and document your findings. A discussion on the value of performing material balances is also provided. Detailed guidance on how to construct flow diagrams and develop written process descriptions follows the material balance discussion.

FLOW DIAGRAM: USE AND CONSTRUCTION TIPS

Performing a walkthrough provides you with an opportunity to become more familiar with each waste generating process and operation. The information and data collected as the result of this effort are most useful when summarized and recorded in the form of a narrative description and as block/process flow diagrams.

The practice of employing flow diagrams to graphically depict unit operations or interrelated steps in manufacturing processes has been around for years. Recently, however, there has been a growing interest in the use of flow diagrams for the purposes of waste tracking, pollution prevention and toxics use reduction planning.

FLOW DIAGRAMS DEFINED

Types of Diagrams

Theodore and McGuinn[6] define four basic types of schematic diagrams that are widely used: (1) the block diagram, (2) graphic flow diagrams ("used most frequently in advertising, company financial reports, and technical reports in which certain features ... are presented in an eye-catching fashion that is both novel and informative"), (3) process flow diagrams, and (4) process piping and instrumentation flow diagrams. Of the four, block diagrams and process flow diagrams are the most useful for hazardous waste tracking purposes.

Block/Process Flow Diagrams

Block diagrams are simple to construct and consist of square and/or rectangular symbols. As described by Theodore and McGuinn, they are "the least descriptive of the schematic diagrams and usually represent a single [step or] unit operation in a plant or an entire section of the plant." By comparison, they say, the process flow diagram is more detailed in that it gives "the basic processing scheme, the basic control concept and the process information from which equipment can be specified and designed." Pojasek and Cali[7] further describe the process flow diagram as "depicting a series of steps through which input materials pass in the course of their transformation into products. For any operation, be it the manufacture of chairs or the maintenance of a pump, there is a functional sequence of events or actions. The process flow diagram clearly illustrates the functional sequence ... of unit operations that lead to the final product or the end of the process."

PROCESS DOCUMENTATION AND RECORDKEEPING

As an important part of your recordkeeping system, block/flow diagrams (1) provide permanent documentation of each waste generating process, (2) demonstrate compliance with federal law by serving as a basis for your waste determination efforts, (3) identify and profile unit operations and material use within a given process, and (4) organize useful information to help you understand and evaluate interrelated steps and the potential for making process improvements. Feedstock materials, direction of work flow, environmental losses and points of waste generation are all essential components of your block/flow diagram (hereafter referred to as flow diagram). In combination with material balances and written process descriptions, flow diagrams help to define material systems and communicate a basic understanding of how each unit operation affects the characteristic(s), composition and quantity of solid and hazardous waste generated.

Level of Detail

Flow diagrams may be drawn to varying levels of detail and arranged in several ways. Organization and level of detail are, however, largely determined by the overall purpose for which the diagram is constructed and by the message(s) to be communicated. For the purposes of waste tracking and cost accounting, simplified flow diagrams should meet the needs of most generators. Before constructing your flow diagram, however, you should first determine the level of detail that will be necessary to meet your individual objectives. In each case, be sure to thoroughly document your findings so that additional data collection efforts will not have to be undertaken again at some future date. The following flow diagram construction tips are offered as general guidance:

- Clearly define process boundaries.[8]

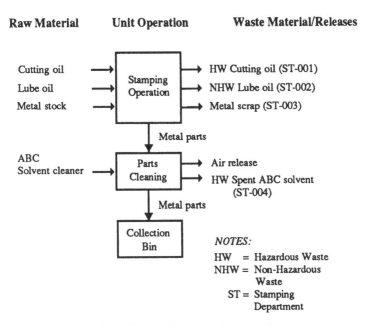

Figure 8. Flow diagram of a typical machine stamping operation.

- Subdivide complex processes into two or more subprocesses where necesary.[9]
- Show all material inputs, direction of work flow, product outputs, chemical releases and waste streams.
- Standardize your use of symbols and keep them simple.
- Mark each solid and hazardous waste stream shown in the flow diagram with its waste identification number.

Detailed guidance on how to construct flow diagrams for waste tracking purposes is provided at the end of this chapter.

Machine Stamping Operation Example

Figure 8 illustrates one method that organizes a flow diagram by input materials, unit operations and wastes generated. In this example, the company mechanically stamps or "cuts" patterns into the surface of smooth metal feedstock. Cutting oils used in the operation are then removed from the work by immersion into a solvent bath. Once free of oil, pieces are collected in a small bin — ready for the next operation. Altogether, three solid wastes are generated in the stamping operation and one as the result of the parts cleaning step; note that solvent vapors are also generated during the parts cleaning operation. Each of the four solid waste streams has been assigned a unique source identifier which allows fully-loaded waste management costs to be tracked directly back to the unit operation (i.e., stamping or parts cleaning operation) that generated the waste; when routinely accounted for, these costs can then be factored into the overall cost for manufacturing a given product. Figure 9 provides a second example of a simple flow diagram where solid and hazardous wastes have been marked with an appropriate waste identification number for tracking and cost accounting purposes. Detailed guidance on off site tracking and fully-loaded waste management cost accounting by point of generation is provided in Chapters 5 and 6.

As a final note, Figures 8 and 9 have been provided for example only. Your diagrams may be more detailed and, at a minimum, should contain the actual chemical names and volumes of each raw material used, released and generated as waste. Also, if other waste materials, such as oily rags or contaminated absorbent, were generated during the stamping operation, for example, then they too would be assigned a waste identification number and shown in the diagram. It is very important that all hazardous and non-hazardous waste streams, air releases and water discharges are accounted for and thoroughly documented. Certain solid wastes and releases, though not regulated as hazardous waste, may represent significant costs and/or potential liability and should, therefore, be tracked and properly managed.

THE MATERIAL BALANCE

Flow diagrams help to map out material systems, and by examining them, you can better visualize material balances, sources of wastes and potential opportunities for process improvement or pollution prevention.

The construction of a complete and accurate process flow diagram represents the first step in the development of a material balance. By definition, the material balance provides an accounting of all chemical inputs and outputs of a process or unit operation. With today's recordkeeping and reporting demands, materials accounting is emerging as a key element in the administration of environmental management programs.

In his article, "Chemical Accounting for the 90s," Wood[11] provides a general overview of the material balance concept:

> In the 1700s, the French chemist, Antoine Lavoisier, weighed the reactants and products of chemical reactions and established the law of conservation of matter in chemical change. This law can be stated as "matter is neither lost nor gained in a chemical reaction." This means that in any chemical reaction, the sum of the weights of the reactants must equal the sum of the weights of the products. For determining this basic principle, Lavoisier has been acclaimed as the father of modern chemistry. Lavoisier could also be called the first chemical accountant.

37

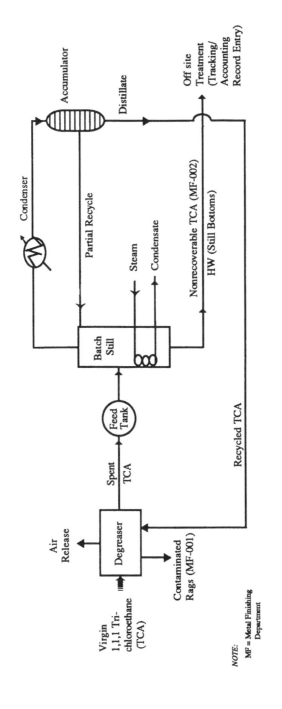

Figure 9. Sample schematic of a batch distillation process.
Adapted from A Compendium of Technologies Used in the Treatment of Hazardous Wastes.[10]

...We might restate Lavoisier's law by saying "whatever you do with chemicals, matter is neither gained nor lost." The most basic principle for a user of hazardous chemicals to apply is [that] the amount of chemicals brought into the facility must eventually equal the amount of chemicals which leave the facility. These amounts would include the chemicals which react to form other chemicals, plus the amounts which leave the facility either in finished products or by-products. Also included must be any amounts which are released into the environment or remain in the facility. The amount of chemical that remains may be assumed to be unused and will be added to new material. This is not always the case, but if the chemical is unused, it must finally leave the facility, possibly as a hazardous waste (Wood, "Chemical Accounting for the 90s." Reprinted with permission of Environmental Waste Management Magazine © 1990).

APPLICATIONS AND EXAMPLES

From an administrative perspective, the material balance is important to your solid and hazardous waste accounting efforts as it enables you to estimate and track off site transfers and material releases to each environmental medium. The material balance also provides you with information needed to perform full cost accounting and assists you in the determination of waste material composition and in the selection of test methods to properly characterize your waste. The U.S. Environmental Protection Agency[12] explains that material balances "allow for quantifying losses or emissions that were previously unaccounted for. They also are useful in the development of (1) a baseline for tracking progress of waste minimization efforts, (2) data to estimate the size and cost of additional equipment and other modifications, and (3) data to evaluate economic performance. The material balance should be made individually for all components that enter and leave the process. When chemical reactions take place in a system, there is an advantage to doing 'elemental balances' for specific chemical elements in a system. Material balances are easier, more meaningful, and more accurate when they are done for individual units, operations, or processes. For this reason, it is important to define the material balance envelope properly. The envelope should be drawn around the specific area of concern, rather than a larger group of areas or the entire facility. An overall material balance for a facility can be constructed from individual unit material balances."

Solvent Operation Case Study

Sometimes a simple comparison between the volume of a material used

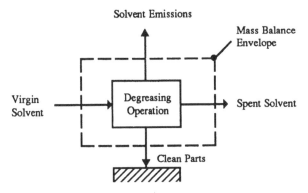

Figure 10. Solvent degreasing flow diagram. Adapted from the EPA publication *Estimating Releases and Waste Treatment Efficiencies.*[13]

and the amount shipped off site as waste will yield an acceptable estimate of release to the environment. Consider Figure 10, for example. In solvent degreasing and drying operations, it is not unusual for companies to experience a 50% to 75% loss in solvent to the surrounding environment (shown in the diagram as Solvent Emissions). Depending upon the cost and volume of virgin solvent purchased, these releases could add up to thousands of dollars in lost funds, as illustrated in the following Case Study.

Case Study: Hypothetical
A widget manufacturer compared the total amount of Freon purchased and used on site to the amount transferred to an off site reclamation facility. In one year the company purchased and used sixteen 55-gallon drums of virgin Freon. By year's end, the company had manifested off site only four drums. The vendor (off site recycler) credited the company $100 for each drum of spent Freon that it received. A simple mass balance calculation [Mass In (raw material) = Mass Out (air emissions + off site transfers)] revealed that the company had lost about 75% or 660 gallons of its Freon to the surrounding environment. Since each drum of virgin Freon cost the company about $2,200, the overall economic loss was substantial — more than $26,000.

Printed Wiring Board Manufacturer Example

In their paper "Waste Minimization Assessments," presented at the AICHE 1990 Summer National Meeting held in San Diego, Higgins and Gemmell[14] provide examples of industry mass balance values for "a general printed wiring board facility." In Figure 11, the facility uses electroplating, electroless plating, and etching operations in the production of 1,300 sq. ft./ day of printed wiring boards. These operations result in the generation of sludge, spent baths, rinse waters and spent copper etchant which is re-

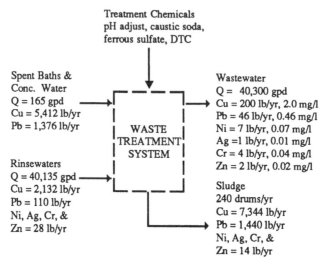

Figure 11. Mass balance for printed wiring board manufacturer's waste treatment operation.[15] *Adapted and reprinted with permission of author.*

claimed (not shown in the diagram). The rinsewaters and spent baths are treated on site. Flow rates (Q represents flow rate and gpd is equivalent to gallons per calendar day) and the amount of metals entering and leaving the board manufacturer's waste treatment system are shown; note that the sum of the quantities of each metal entering the waste treatment system, e.g., 7,544 lb. of copper per year, is equal to the amount leaving it (Mass In = Mass Out).

After performing a material balance, processes and operations take on new meaning. Analysis of the printed wiring board facility data shows, for example, that most of the metals are precipitated out of solution and concentrated in the sludge. The total amount of metal-bearing sludge discarded and associated economic loss is striking — 240 drums per year and $85,000 in lost revenue (1989 dollars). Rather than generate large quantities of sludge or discard valuable metals, companies are far better off investing in pollution prevention methods/technologies that can recover materials at their source. From the examples provided, it's easy to recognize how valuable the material balance is to hazardous waste tracking, pollution prevention and full cost accounting initiatives.

Regulatory Applications

In addition to the uses described above, material balances are sometimes needed to determine the applicability of environmental regulations or meet government reporting requirements.

Table 6 provides examples of RCRA and SARA Title III mass balance-related regulations. SARA, or the Superfund Amendments and Reauthori-

Table 6
Examples of RCRA and SARA Title III Material Balance Regulatory Requirements

Regulation	*Descriptive Language*
RCRA 40 CFR § 261.3	The following mixtures of wastewater and listed hazardous waste are not hazardous waste provided that the generator can demonstrate that the mixture consists of wastewater the discharge of which is subject to regulation under either Section 402 or Section 307(b) of the Clean Water Act and: (A) One or more of the following spent solvents listed in §261.31 - carbon tetrachloride, tetrachloroethylene, trichloroethylene - *provided*, that the maximum total weekly usage of these solvents (other than the amounts that can be demonstrated not to be discharged to wastewater) divided by the average weekly flow of wastewater into the headworks of the facility's wastewater treatment or pre-treatment system does not exceed 1 part per million; or (B) One or more of the following spent solvents listed in §261.31 - methylene chloride, 1,1,1-trichloroethane, chlorobenzene, o-dichlorobenzene, cresols, cresylic acid, nitrobenzene, toluene, methyl ethyl ketone, carbon disulfide, isobutanol, pyridine, spent chlorofluorocarbon solvents - provided that the maximum total weekly usage of these solvents (other than the amounts that can be demonstrated not to be discharged to wastewater) divided by the average weekly flow of wastewater into the headworks of the facility's wastewater treatment or pre-treatment system does not exceed 25 parts per million, or ... [a]
SARA 40 CFR PART 372	For each toxic chemical known by the owner or operator to be manufactured (including imported), processed, or otherwise used in excess of an applicable threshold quantity in §372.25 at its covered facility described in §372.22 for a calendar year, the owner or operator must submit to EPA and to the State in which the facility is located a completed EPA Form R (§372.30). To complete Form R, companies must perform a mass balance on each regulated chemical and report its release to each environmental medium (i.e., air, land and water) including off site transfers. [b]

[a] See 40 CFR §261.3 for complete text.

[b] Under §313(l) of SARA Title III, the EPA Administrator is required to arrange for a mass balance study to be carried out by the National Academy of Sciences using the mass balance information supplied by the regulated community. In §313(l)(4), EPA defines the term "mass balance" as an accumulation of the annual quantities of chemicals transported to a facility, produced at a facility, consumed at a facility, used at a facility, accumulated at a facility, released from a facility, and transported from a facility as a waste or as a commercial product or byproduct or component of a commercial product or byproduct.

zation Act Title III, is also known as the Emergency Planning and Community Right-to-Know Act of 1986. In 40 CFR §261.3, Definition of Hazardous Waste, EPA sets conditions for when mixtures of wastewater and selected solvents are exempt from regulation as hazardous waste. In 40 CFR Part 372, EPA defines when owners and/or operators of facilities that manufacture, process or otherwise use listed chemicals must report mass balance values for chemical releases to each environmental medium. Each example serves to highlight the fact that material balances are not only useful with regard to on site waste tracking, but can also play an important role in environmental compliance.

An Example of How the Material Balance
Concept Can be Used in a Hazardous Waste Determination
40 CFR §261.3, Table 6, states that a mixture of Methyl Ethyl Ketone (MEK) and wastewater is not a hazardous waste if the generator can demonstrate that the mixture consists of MEK and wastewater the discharge of which is subject to regulation under either Section 402 or Section 307(b) of the Clean Water Act and provided that the maximum total weekly usage of MEK (other than the amounts that can be demonstrated not to be discharged to wastewater) divided by the average weekly flow of wastewater into the headworks of the facility's wastewater treatment (WWT) or pre-treatment system (Equation 3.1) does not exceed 25 parts per million (ppm).

The sample problem that follows illustrates how the material balance, based entirely on the generator's knowledge of the waste generating operation, can be used to determine whether a waste mixture is subject to regulation as a hazardous waste.

Problem Statement

Methyl Ethyl Ketone (MEK) is used for its solvent properties to clean certain equipment in a chemical manufacturing facility. Each time the equipment needs to be cleaned (once per week) 45 gallons of MEK is used. Though most of the MEK (30 gallons) that can be captured after cleaning is contained in a 55-gallon drum, some of the chemical volatilizes and some is sent to the company's 1.2 million gallons per week wastewater treatment system. Based on this knowledge of the waste generating operation, determine whether the MEK/wastewater mixture is a RCRA hazardous waste.

Factual Basis for Hazardous Waste Determination:

Generator's Knowledge of Waste Generating Operation
(1) 45 gallons is the maximum total weekly usage of MEK.
(2) 30 gallons of MEK per week is collected in a 55-gallon drum for off site shipment as hazardous waste.

(3) The weekly flow of wastewater to the headworks of the on site treatment facility is 1,200,000 gallons.

(4) MEK is listed in 40 CFR §261.31 and may be exluded from regulation as a hazardous waste if, as a mixture with wastewater, certain conditions are met.

Unknown

(1) Though some MEK volatilizes during the cleaning operation, the exact quantity of MEK that is released to the air is unknown.

Calculation

Based on the language contained in 40 CFR §261.3, the following mass balance equation results:

$$\frac{\text{Max. Total Weekly Usage of MEK - Amount Not Discharged}}{\text{Weekly Flow of Wastewater into WWT Facility Headworks}} \qquad (3.1)$$

Applying the numbers from the problem statement:

$$\frac{45 \text{ Gal. MEK } - 30 \text{ Gal. Spent MEK}}{1,200,000 \text{ Gal. Wastewater}} =$$

$$\frac{15 \text{ Gal. MEK}}{1,200,000 \text{ Gal. Wastewater}}$$

Conversion factors for parts per million calculation (on a weight per weight basis) where:

Density of MEK = 6.71 lb. per gallon (@20°C)

Density of Wastewater = 8.34 lbs. per gallon (@15°C)
(for sample calculation, assume same as raw water)

1 kg = mass of 1 liter of water @4°C

1 kg = 2.2 lb.

1 kg = 10^6 mg

Problem solved using dimensional analysis:

$$
\frac{15 \ \text{Gal. MEK} \times \dfrac{6.71 \ lb.}{1 \ \text{Gal. MEK}} \times \dfrac{1 \ kg}{2.2 \ lb.} \times \dfrac{10^6 \ mg}{kg}}{1{,}200{,}000 \ \text{Gal. Wastewater} \times \dfrac{8.34 \ lb.}{\text{Gal.}} \times \dfrac{1 \ kg}{2.2 \ lb.}} \approx
$$

$$
\frac{10 \ mg}{kg} \sim \frac{10 \ mg}{1} \quad \text{or} \quad \approx \ 10 \ ppm \tag{3.2}
$$

Solution to Hazardous Waste Determination Problem Based on Mass Balance Calculation: The amount of MEK released to the facility's permitted wastewater treatment plant does not cause the MEK/wastewater mixture to be regulated as hazardous waste since the maximum amount of MEK that could possibly enter the headworks of the treatment system divided by the average weekly flow of wastewater does not exceed 25 parts per million (3.2).

SOURCES OF INFORMATION FOR CONSTRUCTING PROCESS FLOW DIAGRAMS

Pojasek and Cali[16] cite two "excellent" sources[17] "that discuss the basics of preparing a process flow diagram." Higgins and Gemmell[18] also report that the Ontario Waste Management Corporation's *Industrial Waste Audit and Reduction Manual* (1987) provides detailed technical examples of flow diagrams, mass balances and evaluation methods. Guidance on how to perform mass balance calculations for Toxic Chemical Release reporting can be found in the manual *Estimating Releases and Waste Treatment Efficiencies for the Toxic Chemical Release Inventory Form* (EPA 560/4-88-002).[19]

The state of Massachusetts[20] has developed *A Practical Guide to Toxics Use Reduction,* which provides detailed guidance on how to create process flow diagrams to comply with state-wide toxic use reduction planning and reporting requirements. In addition to the EPA *Facility Pollution Prevention Guide* cited earlier, examples that illustrate the use of material balances in the performance of "total facility cost" accounting (for waste treatment processes and operations) can be found in the guide *Environmental Pollution Control Alternatives: Reducing Water Pollution Control Costs in the Electroplating Industry* (EPA/625/5-85/016). This manual was prepared jointly by the U.S. Environmental Protection Agency's Industrial Technology Division, Office of Water Regulations and Standards, Office of Water, Washington, D.C. 20460 and the Center for Environmental Research Information, Office of Research Program Management, Office of Research Development, Cincinnati, OH 45268.

DETAILED RECORDKEEPING GUIDANCE

Process Flow Sheets referred to in this chapter can be purchased from a commerical printer/supplier or designed in-house using a computer graphics software program (as noted in Chapter 2, page 17). Figure 12 is one example of a simple process flow sheet that was created using an Aldus Pagemaker® program. The process flow sheet shown in this figure has four notable features indicated by the numbers ❶ through ❹ and as described below. The reverse side of the form, not shown in Figure 12, should be lined and used for recording written process descriptions.

The remaining text of this chapter offers detailed guidance on how to record and organize your data using either custom (Figure 12) or pre-printed forms.

FLOW SHEET HEADING

The Waste Identification Number or Source Identifier should be recorded in the upper right-hand corner (no. ❶, Figure 12) or in the lined space (no. ❷) provided at the top of your process flow sheet. Quite often, an operation or process will generate more than one type of waste; see Figure 8, for example. When this is the case, record each unique source identifier in the space provided; e.g., Figure 8, ST-001/2/3/4. A separate generator waste data sheet is then completed, as outlined in Chapter 4, for each waste stream identified.

The name of the waste generating process, the date that the process review took place, the name of the person performing the review and other important information should also be recorded at the top of the page.

SCHEMATIC CONSTRUCTION

Marked along the upper left-hand margin of the sample Process Flow Sheet are the words "Flow Diagram." This grid area (no. ❸) is used for recording schematics of waste generating processes and operations.

For the purposes of your recordkeeping and accounting system, waste flow diagrams should be concise and easy to understand. Exhaustive detail should not be recorded in the grid area, but rather summarized in narrative form in the lined space on the reverse side of the flow sheet. When there are many steps to a process, additional flow sheets may be needed and should be numbered consecutively.

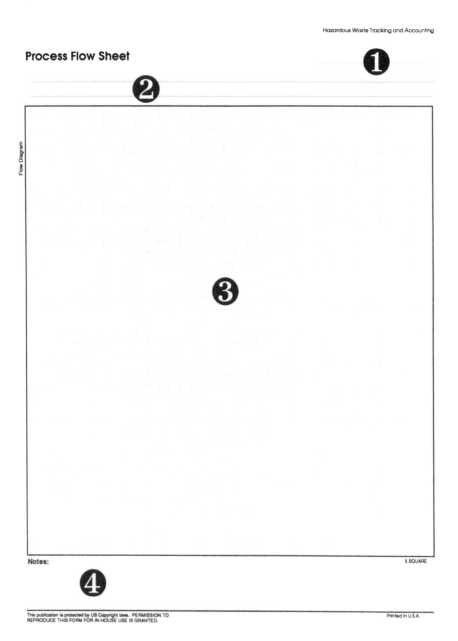

Figure 12. Example of a simple process flow sheet.

For wastes that are generated as the result of a spill or non-process related operation or activity, e.g., MN-003, MN-008 or MN-009 (Figure 6), a short narrative description may be all that is needed. This information can be recorded on the reverse side of the appropriate Solid & Hazardous Waste Index Sheet as discussed in Chapter 2.

Complete one flow sheet for each waste generating process. Clearly define the boundaries of each process. Schematics should show all interrelated unit operations and should be drawn to the level of detail necessary to communicate a basic understanding of chemical, material and work flow.

The following guidelines[20] are intended to assist you in your efforts to construct simple flow diagrams for waste tracking:

1) Draw a box, rectangle, circle or symbol for each individual process unit.

2) Label each unit using either the name of the process (for example, "batch still" — Figure 9) or a process code.

3) Use arrows with labels to show where reagents, chemical additives and other raw materials enter the process.

4) Identify where waste, including treatment residuals, enters or exits a process, by drawing and labeling arrows leading to and from it. Indicate whether waste entering and/or exiting a process is hazardous or nonhazardous by writing HW or NHW. Wherever possible, identify the wastes in general terms in parentheses — for example, "HW (Still Bottoms)."

5) Indicate the point at which residuals that are not managed on site are sent off site.

6) Indicate where wastes originated, including if they came from a system described in another schematic.

7) Indicate locations to which system discharges are routed, including routings to other schematics.

8) Indicate and identify all Air Pollution Control Devices.

In addition to the above guidelines, be sure to mark each waste destined for off site treatment and disposal with its waste identification number as well as mass balance values for each input material. Add any other data or information that you deem appropriate. Notes of special significance can be recorded at the bottom of the flow sheet (no. ❹).

PROCESS/SOURCE DESCRIPTION

General

The reverse side of each flow sheet should be used to record a narrative description of each waste generating process. Developing a written narrative description for each point of generation review may take more time for completion than any other recordkeeping or accounting task. It is very important, however, that a complete and accurate description of each waste generating process, operation and activity is made.

Special attention should be given to detail and the thorough documentation of each finding. Carefully note how raw materials are used and wastes are generated — remember the TCE waste solvent examples referred to in Table 4.

Process descriptions should be adequately detailed so that they can be read and understood by any technically competent person who is unfamiliar with the specifics of your plant's operation.

The narrative should be written in consultation with those who are most familiar with the waste generating process. Only the most pertinent data and information should be recorded. Keep to the facts and boil down the language into its most essential components. Information that is not relevant should be left out. Take special care to create a factual summary that traces the precise sequence of steps defined in the flow diagram.

Specific

Begin the narrative, where appropriate, with a concise profile of the product line. A description of the product's end use, material composition, and rate of production (for example, units produced or processed per unit time) should be recorded. The narrative should then continue with a comprehensive technical review of the operation, process, or activity that results in the generation of waste. Careful attention should be given to record exact process and operational sequences. Facilities, materials, equipment and unit operations should be described. It is sometimes helpful to key the narrative to important steps outlined in the waste flow diagram. Table 7 summarizes the type of information that may be included in your process description.

Narratives describing wastes generated as the result of an operation, incident or, for example, maintenance activity, should be recorded on a Solid & Hazardous Waste Index sheet. The circumstances contributing to and resulting in the generation of these types of waste should be described in detail.

Table 7
Process Description Information

Product Information
Product description including end use, composition, and rate of production.

Raw Materials
Feedstock material summary: chemical name(s), amount, concentration, order of addition, hazards, and losses to the environment.

Summary of Operations
Production/ancillary equipment specifications, layout, design/operating parameters and procedures.
Work flow and methods of material transfer, times of peak or seasonal production, continuous or batch production.
Special sensitivity of product quality or production rate to potential process modification.

Waste Management/Release Data
Points, type(s) and rates of waste generation.
Points at which waste materials are mixed or combined.
Containerization, management, on site/off site treatment and disposal.
Quantitative/qualitative data and information on releases to each environmental medium.
Waste reduction potential and recommendations for improvement.

In all cases, be sure to record the results of material balances and information on chemical releases to each environmental medium (air, land and water). Where available, quantitative data, rates of chemical use/release and analytical results should be recorded and summarized. This data and information will provide a point of departure for other chemical release tracking and reporting efforts.

Where practical, process sheets, operational procedures and other relevant information should be copied and bound together with the appropriate process flow sheet. Clear plastic pocket inserts, available from most office supply stores, are ideal for this purpose.

As a final note, technical reviews should be conducted periodically so that changes in input materials, unit operations or technology can be noted. Keep in mind that Generator Waste Data Sheets, Chapter 4, may also require modification since a process change may affect the characteristic(s) of the waste, regulatory classification, vendor approval status/contract conditions, or treatment and disposal costs.

REFERENCES

1. **40 Code of Federal Regulations** §262.11.
2. **Federal Register,** 50, 6538, 1986.
3. **Federal Register,** 55, 11811, March 29, 1990.
4. **DTI Environmental Programme,** Business and the Environment Unit, *Cutting Your Losses,* Room 1016, Ashdown House, 123 Victoria Street, London SW1EGRB. 1989.
5. **Wigglesworth, D.,** *Profiting from Waste Reduction in Your Small Business, A Guide to Help you Identify, Implement and Evaluate an Industrial Waste Reduction Program,* Alaska Health Project, 431 West 7th Avenue, Suite 101, Anchorage, Alaska 99501. 1988.
6. **Theodore, L. and McGuinn, Y.C.,** *Pollution Prevention,* Van Nostrand Reinhold, 115 Fifth Avenue, New York, New York, 10003. 1992.
7. **Pojasek, R.B. and Cali, L.J.,** Contrasting Approaches to Pollution Prevention Auditing, in *Pollution Prevention Review,* 22 West 21st Street, New York, New York. 10010-6904. Summer 1991.
8. **Goal/QPC,** *The Memory Jogger: A Pocket Guide of Tools for Continuous Improvement,* 13 Branch Street, Methuen, MA 01844. Second Edition. 1988.
9. **op. cit. Pojasek and Cali.**
10. **U.S. Environmental Protection Agency,** *A Compendium of Technologies Used in the Treatment of Hazardous Wastes,* EPA/625/8-87/014, Washington, D.C. September 1987.
11. **Wood, J.L.,** Chemical Accounting for the 90s, in *Environmental Waste Management Magazine,* 243 West Main Street, Kutztown, PA 19530. April 1990.
12. **U.S. Environmental Protection Agency,** *Waste Minimization Opportunity Assessment Manual,* Hazardous Waste Engineering Research Laboratory, Cincinnati, OH 45268. 1988.

    This Manual has been revised and expanded to include "multi-media" pollution prevention. It is now called the *Facility Pollution Prevention Guide* and is published by the Risk Reduction Engineering Laboratory, Office of Research and Development, U.S. Environmental Protection Agency, Cincinnati, Ohio 45268. 1992.
13. **U.S. Environmental Protection Agency,** *Estimating Releases and Waste Treatment Efficiencies for the Toxic Chemical Release Inventory Form,* EPA 560/4-88-002, Office of Pesticides and Toxic Substances, Washington, D.C. 20460. 1988.
14. **Higgins, T.E. and Gemmell, A.S.,** *Waste Minimization Assessments,* Paper presented at the AICHE 1990 Summer National Meeting, San Diego, California. Available through CH2M Hill, P.O. Box 4400, Reston, Virginia. 02209.
15. **ibid.**
16. **op. cit. Pojasek and Cali.**
17. **Ludwig, E.E.,** *Applied Process Design for Chemical and Petrochemical Plants* (Gulf Publishing, 1964) and **O'Donnell, J.P.,** *How Flowsheets Communicate Engineering Information,* Chemical Engineering (McGraw-Hill, 1957).
18. **op. cit. Higgins and Gemmell.**
19. **op. cit. U.S. EPA,** *Estimating Releases and Waste Treatment Efficiencies for the Toxic Chemical Release Inventory Form.*

REFERENCES (Continued)

20. **Office of Technical Assistance for Toxics Use Reduction,** *A Practical Guide to Toxics Use Reduction, Benefiting from TUR at Your Workplace,* Office of Technical Assistance, Executive Office of Environmental Affairs, 100 Cambridge Street, Suite 1904, Boston, MA 02202. 1992.

21. **U.S. Environmental Protection Agency,** *National Survey of Hazardous Waste Generators,* Questionnaire GA, General Facility Information, MD 99-RTI, Research Triangle Park, NC. 1988.

Chapter 4

WASTE CHARACTERIZATION AND DOCUMENTATION
Physical, Chemical, Health and Regulatory Profiles

INTRODUCTION

Obtaining a complete and accurate profile of each waste is an important part of your company's hazardous waste management, tracking and accounting program. In Chapter 3, you performed a series of point of generation reviews. The history concerning how and where wastes are generated was explored in detail. As a result, you are now in a better position to evaluate waste material hazards and chemical composition.

In this chapter, you will identify the chemical constituents and properties of concern for each waste and waste mixture. The chemical makeup of each waste (i.e., chemical type/s and concentration) determines its hazardous nature. Once the hazardous properties are understood, informed management decisions can be made concerning regulatory classification, packaging, storage, worker health and safety, and ultimate treatment and disposal. Knowledge of the physical/chemical properties of each waste allows you to undertake material balance studies, track chemical throughput within your facility and account for total costs associated with product manufacturing and waste generation. As you work on the tasks ahead, it is important to keep in mind that the integrity and long-term success of your waste tracking and cost accounting program is very much dependent upon the quality of your waste characterization data and the thorough documentation of each finding.

GENERATOR WASTE DATA SHEETS

COMMERCIAL APPLICATIONS

Like point of generation indices, waste data sheets have been in use for a number of years. Commercial treatment, storage and disposal facilities, for example, typically require generators to file written technical profiles of each waste prior to material acceptance. The generator usually provides this information and certifies its accuracy on data sheets supplied by the vendor. A "desktop" review of the data is then performed by the TSDF (in conjunction with sample analysis) and the waste is either conditionally accepted or rejected. Though certain information is almost always requested, some data requirements do vary according to facility permit limitations, waste type and method of treatment or disposal.

GENERATOR DATA NEEDS

In addition to qualifying wastes for off site acceptance, generators have their own specific and unique set of data requirements and information needs. Detailed data on the composition of wastes are often needed, for example, to properly characterize wastes, track the fate of chemical constituents, effectively administer on site programs, evaluate material systems, undertake pollution prevention studies (see "Pollution Prevention Example"), ensure worker health and safety and project future needs. To be most effective, this data must not only be of sound quality, but must also be thoroughly documented and clearly recorded. The generator waste data sheet, described below, presents one way to organize physical, chemical, and health hazard data for greater program effectiveness.

Pollution Prevention Example

In Table 8, selected "attributes" and "priority rating criteria" from EPA's Waste Stream Summary work sheet are shown. The work sheet is intended to assist generators in the prioritization of waste streams during pollution prevention assessment studies. In this example, ranking each waste stream for further analysis requires a complete understanding of waste material composition, as well as a good grasp of the environmental, safety and health hazards posed by each waste. In practice, the evaluation of safety/environmental health hazards posed by a waste is sometimes based on laboratory analysis and almost always on the review of available data or literature (see "Waste Determinations and Hazard Analysis" and "Data Sources for Waste Hazard Analyses" in this chapter).

ORGANIZING DATA: WASTE DATA SHEETS

Historically, TSDFs and many of the larger generators have used profile or waste data sheets to catalogue and record summaries of technical data on the origin, composition, properties and classification of each waste. This is a simple practice that can be adopted by generators of all sizes.

Though waste data sheets can be designed to varying levels of detail, "ideally" they should include summary data corresponding to Sections I through VIII of Figures 13 and 14. The organization of your data into a structured format not only improves administrative efficiency, but will also help you to comply (as well as document compliance) with U.S. EPA waste characterization and worker protection/personnel training requirements.

As an information management tool, data sheets can help you meet the day-to-day needs of your waste management program. Listed below are a few ways that waste data sheets can be used by you in the administration of your hazardous waste management program:

Table 8
U.S. EPA Pollution Prevention
Assessment Waste Stream Summary Worksheet [a]

Attribute		Stream No. _____		Stream No. _____	
Waste ID/Name					
Source/Origin					
Component or Property of Concern					
Annual Generation Rate (units ___)					
Overall					
Component(s) of Concern					
Priority Rating Criteria	Wt. (W)	Rating (R)	R x W	Rating (R)	R x W
Waste Hazard					
Safety Hazard					
Potential By-product Recovery					

[a] Adapted from the U.S. EPA *Facility Pollution Prevention Guide,* Office of Research and Development, Washington, D.C. 20460. May 1992.

- As a master record and central information resource; used as a standard reference, for example, for completing government reports, surveys and TSDF waste profile sheets.
- As documentation for hazardous waste determinations.
- As a technical bulletin for hazard communication, personnel training and emergency response operations (discussed in the following section).

The sample generator waste data sheet shown in Figures 13 and 14 has been carefully structured so that technical information and data can be quickly and conveniently located. Each section contains specific data sets that represent the type of information (1) most often needed by generators to administer on site waste management programs, (2) typically used by generators to properly characterize and safely manage wastes, and (3) most frequently requested by off site treatment, storage and disposal facilities as part of the waste acceptance procedure. A discussion on how to construct and use waste data sheets in your facility is provided in the Detailed Recordkeeping Guidance section of this chapter.

COMMUNICATING WASTE MATERIAL HAZARDS

Data sheets can be used to communicate waste material hazards to company employees and others. When workers are thoroughly informed and adequately trained, they are more willing and better able to follow proper health and safety practices when managing hazardous waste. The "Health Hazard Data and Personal Protection" section of the data sheet shown in Figure 14, for example, can be used by generators to record infor-

Hazardous Waste Tracking and Accounting

Generator Waste Data Sheet

_____

Section I – General Waste Description

Description of Waste

Process Generating Waste

Method of Collection/Containerization | Rate of Generation | ☐ One Time ☐ Month ☐ Quarter ☐ Year

Section II – Chemical Composition

Chemical Name (Common Name (s)) | Formula | CAS Number | Concentration

Section III – Physical Properties

☐ Solid ☐ Semi-Solid ☐ Liquid ☐ Gas | Specific Gravity ($H_2O=1$)

☐ Multilayered ☐ Bilayered ☐ Single Phase | pH Range

Free Liquids: ☐ Yes ☐ No   % by volume | Halogenation
| ☐ % Bromine _____   ☐ % Chlorine _____

Flash Point (Method Used) | ☐ % Iodine _____   ☐ % Fluorine _____

Appearance and Odor | BTU/Lb.

Vapor Pressure (mm Hg) | Ash/Dry Weight

Section IV – Reactive Properties

☐ Pyrophoric | ☐ Shock Sensitive

☐ Explosive | ☐ Water Reactive

Printed in U.S.A.

Figure 13. Sample generator waste data sheet (front view).

57

Section V – Metals and Other Components

Arsenic (As)	Cadmium (Cd)	Lead (Pb)	Selenium (Se)
Barium (Ba)	Chromium (Cr⁺)	Mercury (Hg)	Silver (Ag)

Section VI – Health Hazard Data and Personal Protection

Section VII – Spill, Fire and Emergency Response

Section VIII – Regulatory Information

NOTES:

FORM No. 900728/3   Printed in U.S.A.

Figure 14. Sample generator waste data sheet (reverse side).

mation on personal protective equipment requirements, known or suspected health hazards and precautions for safe handling. When combined
with personal instruction, this information can help workers protect themselves during routine operations as well as emergencies involving hazardous waste.

Tracking and documenting chemical/material hazard data in a central
record can also help you meet the hazard communication data submission
requirements of some TSDFs. For example, health hazard data is required
by ENSCO (Environmental Systems Company), Inc. of El Dorado, AR,
prior to material acceptance. In order for ENSCO to pre-qualify a waste for
acceptance, the generator must first submit a completed Waste Material
Data Sheet (supplied by ENSCO). The ENSCO data sheet calls for
generators to provide health and safety information on each waste and to
complete the NFPA (National Fire Protection Association) diamond-
shaped hazard code diagram for health, flammability and reactivity. EN-
SCO also requires[1] generators to submit, where available, "safety sheets,
toxicology reports, TSCA notifications or other information which describes hazards associated with the waste constituents, including material
used in a process [Ch. 3] that might be carried over into the waste stream."

The following case study describes a second TSDF's efforts to ensure
worker health and safety among its employees during waste management
operations.

Case Study

A large commercial TSDF has established the practice of issuing a
"Waste Safety Sheet" to transportation personnel before hazardous waste is
picked up from client generators. The safety sheet serves to inform employees and contractors of proper worker health and safety practices and of the
hazards associated with each waste. The safety sheet contains data on required personal protective equipment, health and fire protection information, reactivity, chemical composition/compatibility and spill response procedures. In addition, the company devised its own in-house "Hazard Code"
warning system which relies on a color-coded hexagon and numerical key
to communicate the degree of hazard for each of six properties: reactivity,
ignitability, corrosivity, acute local toxicity, acute systemic toxicity and
chronic toxicity. Before engaging in any waste management activity, transportation and facility personnel are instructed to read each safety sheet and
take the necessary precautions.

Data Sheets as a Resource for Worker Health and Safety

Personnel Training

Generators like commercial TSD facilities (Case Study above) are faced
with daily decisions concerning worker health and safety. Companies that

generate hazardous waste are responsible for the health and safety of their employees and are obligated to inform them of the hazards of materials that they may come into contact with. Federal regulations, for example, require generators to train facility personnel in "hazardous waste management procedures relevant to the positions in which they are employed (40 CFR § 265.16)." This can mean that, for material handlers and others involved in routine waste management operations, proper instruction must be given on the hazards posed by wastes as well as personal protection measures and measures to be taken in the event of a release or spill. This regulation also requires training that enables personnel to respond effectively to emergency situations such as fires, explosions or unplanned sudden releases of hazardous waste. For those who are charged with the responsibility to train others in hazardous waste management and for those who must respond to emergencies involving hazardous waste, generator waste data sheets provide a practical framework for communicating waste materials hazards.

Transportation Emergency Response Information

Under the Emergency Response Information Communication Standard (49 CFR § 172.602a), generators who offer hazardous waste for transportation are also required to provide emergency response information to carriers that, at a minimum, contains the following information: (1) the basic description and technical name of the hazardous material required by §§ 172.202 and 172.203k, (2) immediate hazards to health, (3) risks of fire or explosion, (4) immediate precautions to be taken in the event of an accident or incident, (5) immediate methods for handling fires, (6) initial methods for handling spills or leaks in the absence of fire, and (7) preliminary first aid measures. Although the U.S. Department of Transportation does not prescribe a specific format, it does state that the information must be presented on a shipping paper, in a document other than a shipping paper (e.g., material safety data sheet), or in a separate document (e.g., an emergency response guidance document) in a manner that cross-references the description of the hazardous material with the emergency response information (49 CFR § 172.602b3). Here again, the generator waste data sheet can assist you in your regulatory compliance efforts and may be used to collect and organize required information.

DATA COLLECTION AND USE

As waste management programs mature, generators usually become more experienced at acquiring good technical data. Sometimes data on the physical and chemical properties of wastes are derived from detailed laboratory analyses, literature reviews, or Material Safety Data Sheets

(MSDS) on constituents of unreacted waste mixtures. Other times, generators rely on their knowledge of processes generating the waste (for example, knowledge of input materials or predicted chemical reactions), vendor waste material profiles, or more typically, a combination of the data sources and methods cited above. And yet, more than 18 years since the passage of RCRA, some generators still lack complete and reliable sets of waste characterization data. Whether your data collection efforts are fully developed or still evolving, waste data sheets can help you organize and structure your recordkeeping and material accounting efforts. Where reliable data and information are needed to effectively administer your hazardous waste management program, it is your sole responsibility, as the generator, to obtain it.

TRACKING VS. TESTING FOR WASTE CHARACTERIZATION

Data sheets should be used to record technical summaries of all available information on the composition, properties and classification of each waste. Ideally, they will serve as a guide in the development of waste material profiles. When completing data sheets, it is not necessary to acquire laboratory data on each and every waste. Current hazardous waste laws, for example, **do not** require laboratory testing for waste determinations. The regulations[2] clearly state that *if the waste is not listed* a generator may determine whether the waste is a hazardous waste by either *(1) testing the waste or (2) applying knowledge of the hazard characteristic of the waste in light of the materials or the processes used.* A well developed tracking program (comprised of documented process reviews, Chapter 3, and the positive identification of waste constituents in mixtures based on generator knowledge, Chapter 4), therefore, can save a company both time and money since costly "waste testing" for the purpose of regulatory classification can be eliminated or reduced where sufficient documentation exists.

DOCUMENTATION AS THE BASIS FOR WASTE CHARACTERIZATION

Frequently, a generator will have sufficient knowledge (supported by written documentation of a waste generating process) that new or additional testing is not necessary. One example of this was given in Chapter 3 under the heading "The Importance of Understanding Your Waste Generating Processes". In that discussion, the importance of understanding how chemicals are used in a process was shown to be a critical factor in waste characterization. The following two examples take this concept one step further for listed hazardous wastes and waste mixtures.

Electroplating Wastewater Treatment Sludge (F006) Example

Table 9 was assembled by combining information from 40 CFR §261.11 (Criteria for Listing Hazardous Waste), §261.31 (Hazardous Waste from Non-specific Sources), §261.32 (Hazardous Waste from Specific Sources) and Appendix VII (Basis for Listing Hazardous Waste). It presents four types of RCRA listed hazardous wastes along with their EPA-assigned hazard codes, basis for listing and specific hazardous constituents. The table shows that once a listed hazardous waste is properly classified, it is possible to gain insight into the hazardous properties of a waste (without extensive laboratory testing) by using EPA's rationale or "basis for listing" and Appendix VII as a starting point.

EPA's basis for listing waste water treatment sludge from electroplating operations (F006 Hazardous Waste, Table 9) as a toxic waste (T), for example, is that substances contained in the waste have been shown in scientific studies to be toxic (having an adverse impact on a biologic system), carcinogenic (capable of causing or inducing cancer), mutagenic (inducing a change in hereditary material, DNA) or teratogenic (causing developmental malformations) in humans or in other forms of life. This knowledge, coupled with physical, chemical and toxicological data on the hazardous constituents present, provides the generator with an information base on which to build a waste profile. Though four hazardous constituents are shown in Table 9 (with reference to the generic F006 listing), a facility's wastewater treatment sludge may contain one, none or a combination of two or more of the noted constituents depending upon the specific electroplating process(es) employed. In this example, the generator, based on knowledge of the plating chemicals used, would be able to determine the identity of hazardous constituents potentially present in the sludge.

Laboratory Waste Solvent Example

A second example of how a generator's knowledge can be used in a waste determination, in lieu of instrumental analysis, is the case where a laboratory routinely generates mixed waste solvents (see Case Study below). In this example, the laboratory keeps a running log, Figure 15, on the amount and type of spent solvent generated and added to a waste collection container. When the container is full, a simple tally by chemical type provides the basis for waste characterization. In this case, "waste testing", or instrumental analysis would not be required to determine material composition since the history of the mixture is well documented. Furthermore, the generator's knowledge of the waste, combined with published data available in standard chemical handbooks/data bases (see Table 10 for sample listing) or Material Safety Data Sheets, may be sufficient to allow the investigator to properly characterize the waste, understand potential hazards and determine appropriate management

Table 9. Examples of RCRA Listed Hazardous Waste, Hazardous Constituents, and EPA Basis for Listing. [a]

Hazardous Waste No.	Hazardous Waste	Hazard Code [b]	Code Name	Basis for Listing [c]	40 CFR Part 261 Appendix VII Hazardous Constituents
F006	Wastewater treatment sludges from electroplating operations (exceptions cited in §261.31)	T	Toxic Waste	Substances shown in scientific studies to have toxic, carcinogenic, mutagenic or teratogenic effects on humans or other life forms	Cadmium, Cr^{+6}, Nickel, Cyanide (complexed)
F007	Spent cyanide plating bath solutions from electroplating operations	R,T	Reactive & Toxic Waste	Liquid waste meeting the characteristic of reactivity, 40 CFR §261.23, & Toxic listing criteria	Cyanide (salts)
F020	Wastes from the production or manufacturing use of tri- or tetrachlorophenol, or of intermediates used to produce their pesticide derivatives (exceptions cited in §261.31)	H	Acute Hazardous Waste	Fatal to humans at low doses or has been shown in studies to have an oral LD 50 toxicity (rat) of less than 50 mg/kg, an inhalation LC 50 toxicity (rat) of less than 2 mg/l, or a dermal LD 50 toxicity (rabbit) of less than 200 mg/kg or is otherwise hazardous per 40 CFR §261.11 (a) (2)	Tetra- and pentachloro-dibenzo-*p*-dioxins; tetra- and pentachloro-dibenzofurans; tri- and tetrachlorophenols and derivatives cited in Append. VII
K007	Wastewater treatment sludge from the production of iron blue pigments	T	Toxic Waste	Same as for F006 above	Cyanide (complexed), hexavalent chromium

a Source: Title 40 Code of Federal Regulations Part 261 (§261.21-24, §261.31-32 and Appendix VII).

b A waste is listed in 40CFR Subpart D (Lists of Hazardous Waste) based upon one or more of the following criteria: it is Ignitable (I), Toxic (T), Reactive (R), Acutely Hazardous (H), Corrosive (C), or it meets the Toxicity Characteristic (E). Appendix VII of 40 CFR Part 261 identifies the constituent which caused the Administrator to list the waste.

c Criteria for toxic, reactive and acute hazardous waste listings cited. Hazardous waste examples may meet one or more of the criteria.

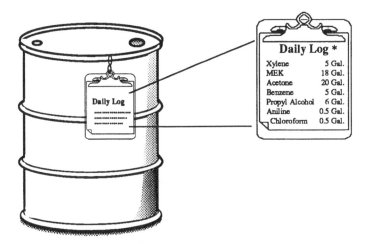

Figure 15. Laboratory solvent disposal drum with clip board for data collection.

methods — refer also to "Lessons from the Lab Solvent Case Study" and "Waste Determination/Hazard Analysis" that follow.

Case Study (Hypothetical)

A research laboratory generates four 55-gallon drums of waste solvent per year — or one every 90 days. Since laboratory staff conduct a wide range of experiments, the types and quantities of solvents contained in a given mixture (55 gallons in volume) can vary over time. Each time a 55-gallon drum of mixed solvents is prepared for shipment, the contents must be fully characterized and the company must obtain preapproval from the prospective TSDF. Since the commingling of incompatible wastes is prohibited under RCRA, staff (1) are careful to mark each small container (i.e., bottle) of spent solvent with a special tracking label (Figure 16), (2) record the same data that appears on the label in a laboratory notebook for future accounting purposes, and (3) identify potential hazardous reactions (through a literature search or appropriate testing — see, for example, *Design and Development of a Hazardous Waste Reactivity Testing Protocol,* Table 10) that could occur with other chemicals prior to consolidation.

Once satisfied that a hazardous reaction will not occur, the spent solvent contained in each bottle is poured off into a 55-gallon drum within which it is commingled with other compatible materials. A chemical profile of one

---

* The mixture shown in Figure 15 is presented for discussion. In practice, waste stream segregation is key to cost containment. Best management practice in this case would be to have separate containers for halogenated and non-halogenated solvents. The presence of even the smallest amount of a halogenated solvent in an otherwise non-halogenated mixture can result in increased disposal costs for the entire 55-gallon volume.

Table 10
Sources of Information
The following lists provide sources of information and data
on chemicals, chemical properties and associated hazards.

Publications for Hazard Identification

Meyer, E., *Chemistry of Hazardous Materials*, Prentice-Hall, Inc., Englewood Cliffs,
New Jersey 07632. Second Edition. 1989.

National Fire Protection Association, *Fire Protection Guide on Hazardous Materials*, Batterymarch Park, Quincy, MA 02269. Tenth Edition. 1991.

Sax, N.I. and R.J. Lewis, Sr., *Dangerous Properties of Industrial Materials*, Van
Nostrand Reinhold Company, 115 Fifth Avenue, New York, New York 10003. Eighth
Edition. 1992.

Sax, N.I. and R.J. Lewis, Sr., eds., *Hawley's Condensed Chemical Dictionary*, Van
Nostrand Reinhold Company, 115 Fifth Avenue, New York, New York 10003. Eleventh Edition. 1987.

Sittig, M., *Handbook of Toxic and Hazardous Chemicals and Carcinogens*, Noyes
Publications, Mill Road, Park Ridge, New Jersey 07656. 1985.

U.S. Department of Health and Human Services, *Pocket Guide to Chemical
Hazards*, National Institute for Occupational Safety and Health, U.S. Government
Printing Office, Superintendent of Documents, Washington, D.C. 20402. June 1990.

U.S. Department of Health and Human Services, *Registry of Toxic Effects of
Chemical Substances (RTECS)*, National Institute for Occupational Safety and Health,
Public Health Service, Centers for Disease Control, Cincinnati, Ohio 45226. RTECS
is now out of print; the 1985-86 edition was the last paper copy available. RTECS data
may be accessed electronically through the (1) National Library of Medicine, (2)
Chemical Information System, (3) Chemical Abstracts Service Science and Technology Network, (4) European Space Agency Information Retrieval System, or (5)
DIALOG.

U.S. Department of Health and Human Services, *Toxicological Profiles* and *Case
Studies in Environmental Medicine*, Agency for Toxic Substances and Disease Registry, 1600 Clifton Road, N.E. (E-29), Atlanta, Georgia 30333.

U.S. Environmental Protection Agency, *Design and Development of a Hazardous
Waste Reactivity Testing Protocol*, Solid and Hazardous Waste Research Division,
Municipal Environmental Research Laboratory, Cincinnati, Ohio 45268. This publication is available from the National Technical Information Service, 5285 Port Royal
Road, Springfield, VA 22161.

Continued

Table 10 (Continued)
Bibliographic Data Bases [3]

Bibliographic Retrieval Services (BRS), 1200 Route 7, Latham, NY 12110.
File Name: Biosis Previews, CA Search, Medlars, NTIS, Hazardline, American Chemical Society Journal, Excerpta Medica, IRCS Medical Science Journal, Pre-Med, Intl. Pharmaceutical Abstracts, Paper Chem.

Chemical Information System ICI (ICIS), Bureau of National Affairs, 1133 15th Street, NW, Suite 300, Washington, D.C. 20005.
File Name: Structure and Nomenclature Search System (SANSS), Acute Toxicity (RTECS), Clinical Toxicology of Commercial Products, Oil and Hazardous Materials Technical Assistance Data System, CCRIS, CESARS.

Lockheed-DIALOG, Information Service, Inc., 3460 Hillview Avenue, Palo Alto, CA 94304.
File Name: Biosis Prev. Files, CA Search Files, CAB Abstracts, Chemical Exposure, Chemname, Chemsis Files, Chemzero, Embase Files, Environmental Bibliographies, Enviroline, Federal Research in Progress, IRL Life Science Collection, NTIS, Occupational Safety and Health (NIOSH), Paper Chem.

National Library of Medicine, Department of Health and Human Services, Public Health Service, 8600 Rockville Pike, Bethesda, MD 20894.
File Name: Hazardous Substances Data Bank (NSDB), Medline files, Toxline Files, Cancerlit, RTECS, Chemline.

Occupational Health Services, 400 Plaza Drive, Secausus, NJ 07094.
File Name: MSDS, Hazardline.

SDC-Orbit, SDC Information Service, 2500 Colorado Avenue, Santa Monica, CA 90406.
File name: CAS Files, Chemdex, 2, 3, NTIS.

---

drum is shown in Figure 15. Since specific chemical names and corresponding waste volumes are known, a waste characterization/worker health and safety summary table (Table 11) can be prepared by simply referring to the published literature ("Publications for Hazard Identification", Table 10 — Sax, *Dangerous Properties of Industrial Materials;* Sittig, *Handbook of Toxic and Hazardous Chemicals and Carcinogens;* and Sax, *Hawley's Condensed Chemical Dictionary*).

Lessons from the Lab Solvent Case Study
The preceding case study shows that laboratory staff had in place a comprehensive solvent waste tracking system — consisting of container labels, notebook recordkeeping and a daily log (Figure 15). This tracking system allowed the generator, based solely on experiential knowledge, to simply and accurately determine the composition of the waste mixture

```
┌─────────────────────────────────────────────────────────────┐
│ HAZARDOUS WASTE                                               │
│                                                               │
│ Chemical Name:             Methyl Ethyl Ketone                │
│ Container Type:            Glass Bottle                       │
│ Material Quantity:         1 Liter                            │
│ Tracking Info:             Research Lab 101                    │
│ Accumulation Start Date:   September 5, 1995                  │
│ Other Information:         Used for solvent properties in     │
│                            organic intermediate experiments   │
│                                                               │
│ SQ Generator, Inc.                                            │
│ Somewhere, New England 01234                                  │
└─────────────────────────────────────────────────────────────┘
```

Figure 16. Sample label used for tracking small containers* of lab waste prior to mixing with other compatible materials (in a 55-gallon drum) for off site shipment.

without costly analytical testing. Since each chemical component was known, along with information on its prior use (for example, in Figure 16 MEK is shown to have been used for its solvent properties and is therefore classified as an F005 "RCRA Listed Waste" in Table 11) and relative quantity, laboratory staff were in a good position to characterize the waste for on site management as well as for off site transportation and treatment. Based on the generator's knowledge of the waste, standard reference materials could then be used to investigate potential chemical hazards and compile supportive data for waste characterization (Table 11).

Waste Determinations and Hazard Analysis

By understanding the history of how chemicals are first used, waste mixtures can be properly classified in accordance with RCRA regulations. Since the identity and volume of each chemical in the preceding case study were known, standard reference books could be used to determine material hazards. In Table 11, for example, two principal hazards — Health and Fire — were investigated.

For this discussion, health hazard data have been subdivided into two columns representing two types of effects — carcinogenic (nonthreshold) and other toxic (threshold) effects. One observation that can be made upon review of the data in Table 11 is that chemicals can present a range of human health risks under varying conditions of exposure. As a general rule, potential adverse human health effects associated with chemical usage are not only dependent upon material hazard, but are directly related to the magnitude, frequency and duration of exposure.

---

* In practice, once bottles are emptied, they should be triple rinsed, labels should be removed or defaced, with black marker (indelible ink) for example, and glass sent to a recycler if container cannot be reused.

Table 11. Summary Table — Waste Determination and Hazard Analysis Based on Generator Knowledge.

Chemical	RCRA Listed Waste				Characteristic Waste		Health Hazard		Fire Hazard [d]	
	F003	U002	F005	U012	D001	D022	Carcinogenic [b]	Toxic [c]	Combustible	Flammable
Xylene	•							•		•
Methyl Ethyl Ketone			•					•		•
Acetone [a]		•						•		•
Benzene			•				•	•		•
Propyl Alcohol					•			•		•
Aniline [a]				•			•	•	•	
Chloroform						•	•	•		

[a] Discarded as unused product; all other solvents discarded when spent.

[b] Benzene: Confirmed human carcinogen producing myeloid leukemia, Hodgkin's disease and lymphomas by inhalation; Aniline: Suspected carcinogen with experimental neoplastigenic data; Chloroform: Confirmed carcinogen (Sax et al. 1992, Table 10).

[c] Xylene: Acute exposure to xylene vapor may cause central nervous system depression and minor reversible effects upon liver and kidneys. At very high concentrations, vapors may cause pulmonary edema, anorexia, nausea, vomiting, and abdominal pain; Methyl Ethyl Ketone: Attacks central nervous system and lungs. Produces irritation of eyes and nose, headaches, dizziness and vomiting; Acetone: systemic effects by ingestion include coma, kidney damage and metabolic changes. Effects by inhalation include changes in EEG, changes in carbohydrate metabolism, nasal effects, conjunctiva irritation, respiratory system effects, nausea, vomiting and muscle weakness (Sax et al. 1992, Table 10); Benzene: Acute exposure results in central nervous system depression. Headache, dizziness, nausea, convulsions, coma and death may result. Death has occurred from large acute exposure or as a result of ventricular fibrillation; Propyl Alcohol: Can produce mild central nervous system depression; Aniline: Absorption from inhalation or skin absorption causes anoxia. If treatment is not given promptly, death can occur; Chloroform: Potent anesthetic. Exposure may cause lassitude, digestive disturbances, dizziness, mental dullness, and coma (Sittig 1985, Table 10).

[d] U.S. DOT — Combustible liquid: flash point >141°F to < 200°F; Flammable liquid: flash point ≤ 141°F. Data used to classify solvents taken from *Hawley's Condensed Chemical Dictionary*, Table 10.

Benzene, for example, is regulated by the U.S. EPA as a Class A Human Carcinogen. This "weight of evidence" classification is based on the fact that there is sufficient evidence from epidemiological studies to support a causal association between exposure to benzene and human cancer. Retrospective epidemiologic studies on workers employed in the shoe industry, rubber product and chemical manufacturing industries, as well as gavage and inhalation studies in rats and mice support this finding.[4] Although the primary acute toxic manifestation of benzene is central nervous system depression, the weight of evidence suggests that repeated exposure to benzene can result in leukemia.[5]

Since the effects of both acute and chronic exposure* to chemical wastes can be severe (e.g., resulting in incapacitating illness or even death), generators need to develop health and safety programs that protect against exposure over a wide range of waste management scenarios. The type of information and data discussed in this section should be evaluated and communicated to workers who are at potential risk during hazardous waste management operations.

The Fire Hazard column, Table 11, shows that most of the chemicals contained in the sample mixture have a flash point — defined as the minimum temperature at which a liquid gives off vapor within a test vessel in sufficient concentration to form an ignitable mixture with air near the surface of the liquid (NFPA, Table 10) — of less than 100°F. In consideration of the relative volumes of each chemical present in the 55-gallon drum, one could reason that (based on published data) the mixture meets the U.S. DOT definition of a Flammable Liquid.

In summary, the knowledge gained from a hazard assessment and waste tracking effort of this type can be used to develop waste material profiles for chemical mixtures at minimal cost. Once complete, waste data sheets can be used by hazardous waste managers as documentation for waste determinations and as a central resource for worker health and safety training and emergency response efforts required by law. Though this was a hypothetical example, in practice the generation of hazardous waste from laboratories should be minimized where practicable.

Data Sources for Waste Hazard Analyses

References listed in Table 10 will assist you in the characterization of wastes and in the evaluation of human health and safety hazards. Several of these references were cited as data sources for waste material hazard evaluations in the preceding hypothetical case study.

---

* In the field of toxicology, the term acute exposure refers to those exposures that last less than 24 hours in duration. Chronic exposures, on the other hand, are repeated exposures which generally occur over a period of more than 3 months.

Though *Hawley's Condensed Chemical Dictionary*, Sax's *Dangerous Properties of Industrial Materials*, Sittig's *Handbook of Toxic and Hazardous Chemicals and Carcinogens* and the NFPA's *Fire Protection Guide on Hazardous Materials* are widely used by hazardous waste generators, three additional sources of data and information can also be useful when conducting detailed hazard assessments:

Registry of Toxic Effects of Chemical Substances (RTECS)

RTECS is a database that is maintained by the National Institute for Occupational Safety and Health under the authority of the Occupational Safety and Health Act of 1970, as amended. This database contains more than 118,000 chemicals which are known to be toxic. Toxicity data have been extracted from the scientific literature and are presented for six categories: (1) primary irritation, (2) mutagenic effects, (3) reproductive effects, (4) tumorigenic effects, (5) acute toxicity, and (6) other multiple dose data. RTECS is updated quarterly and is available on magnetic computer tape, in microfiche form, or may be accessed electronically from five different sources as shown in Table 10.

In addition to its value when conducting hazard assessments, RTECS may also be an important resource for generators who are located in states that require oral rat LD_{50} (the dose of a chemical that is lethal to 50% of the experimental animal population), primary irritation or other unique data for hazardous waste classification. In comparison to standard handbooks, RTECS covers more chemicals and is updated more frequently. For companies with overseas operations, RTECS also contains information on international standards and regulations. A sample printout of selected data for benzene is provided in Figure 17.

Toxicological Profiles

The Administrator of the Agency for Toxic Substances and Disease Registry (ATSDR) is required to prepare toxicological profiles for each substance listed on the Superfund National Priorities List pursuant to the Superfund Amendments and Reauthorization Act of 1986. The revised list, published in the 28 October 1992 Federal Register, contains 275 hazardous substances that may be present at hazardous waste disposal sites. Each profile summarizes the toxicological and adverse health effects of the subject chemical.

Toxicological profiles may be used by safety/health professionals and environmental managers to obtain a more complete understanding of the hazardous properties of waste materials generated and/or managed. The information presented can assist trainers and those responsible for waste hazard communication in their efforts to respond to questions and inform employees of chemical hazards. An outline of a standard ATSDR toxicological profile table of contents is provided in Table 12.

```
1/9/1
DIALOG(R)File 336:Reg Tox Eff Chem Sub
(c) format only 1994 Dialog Info.Svcs. All rts. reserv.

018921          RTECS Number: CY1400000
Substance Name: BENZENE
CAS Registry Number: 71-43-2    Molecular Formula: C6H6   Molecular Weight:
78.12
Synonyms: (6)ANNULENE ; BENZEEN (Dutch) ; BENZINE (Polish) ; BENZENE
   (ACGIH,DOT,OSHA) ; BENZIN (OBS.) ; BENZINE (OBS.) ; BENZOL ; BENZOLE ;
   BENZOLENE ; BENZOLO (Italian) ; BICARBURET of HYDROGEN ; CARBON OIL ;
   COAL NAPHTHA ; CYCLOHEXATRIENE ; FENZEN (Czech) ; MINERAL NAPHTHA ;
   MOTOR BENZOL ; NCI-C55276 ; NITRATION BENZENE ; PHENE ; PHENYL HYDRIDE
   ; PYROBENZOL ; PYROBENZOLE ; RCRA WASTE NUMBER U019 ; UN1114 (DOT)
Compound Class: Agricultural Chemical; Tumorigen; Drug; Mutagen;
   Reproductive Effector; Primary Irritant
Wiswesser Line Notation: RH
Record Date: 9406

IRRITATION DATA:
   Skin  Rabbit     15   mg/24H open  Mild  AIHAAP   23,95,62
   Skin  Rabbit     20   mg/24H Moderate  85JCAE  -,25,86
   Eye   Rabbit     88   mg Moderate  AMIHAB  14,387,56
   Eye   Rabbit      2   mg/24H Severe  85JCAE  -,25,86
MUTATION DATA:
   Microsomal mutagenicity assay  Salmonella typhimurium      10  ppm
       EVHPAZ  82,81,89
   Specific locus test  Drosophila melanogaster    Oral   11250  umol/L
       PMRSDJ  5,325,85
   Sex chromosome loss and nondisjunction  Drosophila melanogaster    Oral
```

Figure 17. RTECS (Registry of Toxic Effects of Chemical Substances) printout on benzene. [6] The complete file for benzene is nine pages in length and includes data on irritation, mutagenic, reproductive, tumorigenic and other toxic effects. A key to journal references is contained in each report; e.g., AIHAAP (noted in Skin Rabbit Irritation Data, above) — American Industrial Hygiene Association Journal, 475 Wolf Ledges Pkwy., Akron, OH V.1-1960. Information on scientific reviews, national/international standards and regulations, criteria documents and NTP/NIOSH/EPA program status reports is also presented.

```
REPRODUCTIVE EFFECTS DATA:
  Female fertility index   Inhalation  Rat  TCLo   670  mg/m3/24H  15D
    pre/1-22D preg  HYSAAV  33(1-3),327,68
  Biochemical and metabolic   Inhalation  Rat  TCLo   56600  ug/m3/24H
    1-22D preg  HYSAAV  33(7-9),112,68
  Extra embryonic structures ;Fetotoxicity (except death)  Inhalation  Rat
    TCLo   50  ppm/24H  7-14D preg  JHEMA2  24,363,80

TUMORIGENIC EFFECTS DATA:
  Carcinogenic by RTECS criteria ; Leukemia ; Thrombocytopenia ;
    Inhalation  Man  TCLo   200  mg/m3/78W-I  EJCAAH  7,83,71
  Carcinogenic by RTECS criteria ; Leukemia ;  Inhalation  Human  TCLo   10
    ppm/8H/10Y-I  TRBMAV  37,153,78
  Carcinogenic by RTECS criteria ; Endocrine--Tumors ; Leukemia ;  Oral
    Rat  TDLo   52  gm/kg/52W-I  MELAAD  70,352,79

TOXICITY EFFECTS DATA:
* Inhalation  Human  LCLo    2  pph/5M   TABIA2  3,231,33
* Oral  Man  LDLo   50  mg/kg  YAKUD5  22,883,80
  Blood--Other changes ;Body temperature increase   Inhalation  Man  TCLo
    150  ppm/1Y-I  BLUTA9  28,293,74
  Somnolence (general depressed activity) ;Nausea or vomiting ;Dermatitis,
    other (after systemic exposure)  Inhalation  Human  TCLo   100  ppm
    INMEAF  17,199,48

REVIEWS:
  ACGIH  TLV-SUSPECTED HUMAN CARCINOGEN   85INA8  6,108,91
  ACGIH  TLV-TWA 10 ppm   85INA8  6,108,91
  IARC  CANCER REVIEW:HUMAN LIMITED EVIDENCE   IMEMDT  7,203,74
  IARC  CANCER REVIEW:ANIMAL SUFFICIENT EVIDENCE   IMSUDL  7,120,87
  IARC  CANCER REVIEW:ANIMAL LIMITED EVIDENCE   IMEMDT  29,93,82
  IARC  CANCER REVIEW:HUMAN SUFFICIENT EVIDENCE   IMEMDT  29,93,82

STANDARDS AND REGULATIONS:
  DOT-HAZARD 3; LABEL:FLAMMABLE LIQUID  CFRGBR  49,172.101,92
  MSHA STANDARD air-CL 25 ppm (80 mg/m3) (skin)  DTLVS*  3,22,71
  OSHA cancer hazard  FEREAC  52,34460,87
```

Figure 17. RTECS printout on benzene (Continued).

Table 12
Contents of ATSDR Toxicological Profile: Toluene Example [a]

Section	Comments
Public Health Statement	This section provides clear and concise answers to the questions: What is toluene? What happens to toluene when it enters my body? What happens to toluene when it enters the environment? How might I be exposed to toluene? How can toluene effect my health? Is there a medical test to determine whether I have been exposed to toluene? What recommendations has the Federal government made to protect human health? Where can I get more information?
Health Effects	The Health Effects section discusses adverse effects following inhalation, oral and dermal exposure. Health effects discussed include: death, systemic effects, immunologic effects, neurological effects, reproductive effects, developmental effects, genotoxic effects and cancer.
Toxicokinetics [b] & Other Data	Toxicokinetic data is given on the absorption, distribution, metabolism, excretion and mechanisms of action.
	Additional information is given on the relevance to public health, biomarkers of exposure and effect, interactions with other chemicals, populations that are unusually susceptible, methods for reducing toxic effects, and adequacy of the data base.
Chemical & Physical Information	Data on chemical identity, physical and chemical properties are provided.
Production, Import/Export, Use and Disposal	Historical and recent domestic production, import/export, use and disposal data are presented.
Human Exposure Potential	This section provides information on releases to the environment, environmental fate mechanisms, levels monitored or estimated in the environment, general population and occupational exposure, populations with potentially high exposures and adequacy of the data base.
Analytical Methods	Analytical methods for biological materials, environmental samples and information on the adequacy of the data base are presented.

Continued

Table 12 (Continued)
Contents of ATSDR Toxicological Profile: Toluene Example

Section	Comments
Regulations and Advisories	In this section, ATSDR provides information on national/state regulations and guidelines. EPA air, water and hazardous waste regulatory requirements are listed. OSHA, NIOSH, ACGIH and NAS regulatory and guideline thresholds are given.[c]

[a] Source: Toxicological Profile for Toluene (Update), Agency for Toxic Substances Disease Registry (ATSDR), Division of Toxicology/Toxicology Information Branch, U.S. Department of Health and Humna Services, 1600 Clifton Road NE, E-29, Atlanta, Georgia 30333. May 1994.

[b] The term toxicokinetics refers to the changes that a toxic substance goes through in the body over time. It involves absorption, distribution, biotransformation and excretion of the toxicant. [7]

[c] OSHA — Occupational Safety and Health Administration; NIOSH — National Institute for Occupational Safety and Health; ACGIH — American Conference of Governmental Industrial Hygienists; NAS — National Academy of Sciences.

Toxicological profiles are available from the U.S. Department of Commerce, National Technical Information Service, 5285 Port Royal Road, Springfield, VA 22161.

Integrated Risk Information System (IRIS)

IRIS provides up-to-date health hazard assessment data for over 600 chemical and physical agents. The data base was developed by the U.S. EPA in 1985 and is continually revised by EPA work group scientists as new information becomes available. Though its primary intent is to provide high quality scientific data to EPA staff for use in the performance of risk assessments, generators of hazardous waste can access the data base for current, peer reviewed information on the health effects and regulatory status of chemical constituents that may be present in their waste. For example, while the latest editions of standard chemical handbooks may indicate that tetrachloroethylene (PCE) is carcinogenic, current IRIS files show that an EPA work group is in the process of reevaluating PCE's potential to cause cancer; as of February 1995, no definitive conclusion had been reached. The identification of new chemicals as potential human carcinogens and changes in EPA's classification of existing chemicals (i.e., Class A: human carcinogen; Class B: probable human carcinogen; Class C: possible human carcinogen; Class D: not classified; and Class E: no evidence of carcinogenicity to humans) is provided in IRIS along with a detailed scientific justification. IRIS contains information on acute and chronic health effects resulting from chemical exposure, drinking water health advisories, organoleptic properties, treatment of chemicals in water, U.S. EPA regulatory actions (under the Safe Drinking Water Act, the Clean Water Act, the Resource Conservation and Recovery Act, CERCLA, and the Toxic Substances Control Act), and supplementary data.

Information on IRIS can be obtained from the Risk Information Hotline 513/569-7159 managed by Labat-Anderson, Inc. or by writing the U.S. Environmental Protection Agency, Office of Research and Development, Environmental Criteria and Assessment Office, Cincinnati, Ohio 45268. IRIS is updated monthly and can be accessed from your U.S. EPA regional library, or electronically through DIALACOM, Inc. (Federal Systems Division, 6120 Executive Blvd., Suite 150, Rockville, MD 20852) or the National Library of Medicine's TOXNET system. The National Technical Information Service also provides IRIS on diskette with quarterly updates.

As a final note, all sources of information listed in Table 10 and discussed above are secondary data sources. In most cases, these data sources will refer you to original studies in the primary scientific literature. In the case of RTECS, no attempt has been made to assess the quality or integrity of the original experiments. In other cases, some level of screening has been performed by the responsible agency or publisher. In the case of IRIS and ATSDR, materials have been subjected to a rigorous peer review.

Sample toxicity profiles for toluene and dichloromethane are provided in Appendix C. These profiles summarize available data and were prepared using IRIS and ATSDR as primary reference sources; other secondary data sources were used to a lesser extent as noted. Each profile demonstrates how secondary data sources can be used by environmental, safety and health professionals to develop written summaries that lead to a general understanding of the hazards associated with chemicals and chemical wastes.

LABORATORY TESTING AS THE BASIS FOR WASTE CHARACTERIZATION

In many cases, laboratory testing for waste characterization or for other purposes (Table 13) will be necessary. Based upon the information acquired in Chapter 3, you are now in a strong position to determine analytical needs. Knowledge of waste generating processes and input materials used provides good insight into waste material composition and hazard characteristics to be expected.

Before you begin any sampling effort, however, take the time to thoroughly *think through* your objective(s). You should not only consider the type of process generating the waste, but also the purpose(s) for which you are conducting the analysis. Consider, for example, the case of the printed

Table 13
Examples of Potential Uses for Laboratory
Data on Waste Materials

- Classify wastes according to state and federal law
- Perform material balance studies
- Properly treat or dispose of hazardous waste
- Assess the appropriateness of waste reduction methods
  and technologies
- Select personal protective equipment
- Develop safe handling and storage procedures
- Mark, label and properly package containers of hazardous
  waste in accordance with the U.S. Department of Transportation's
  Hazardous Materials Regulations

wiring board manufacturer presented in Figure 11. In that example, the company wanted to know the *total* amount of metals (Cu, Pb, Ni, Ag, Cr, and Zn) present in each waste stream so that waste reduction methods could be appropriately evaluated. To get a clear picture of total metal losses, an elemental balance was performed and metal concentration values were converted into annualized waste generation rates. If, by contrast, the company wanted to determine the relative mobility of each metal in a landfill environment, representative samples of the wastewater treatment sludge would have to be analyzed for *extractable* (rather than total) metal content. By carefully considering the history of each waste and the purpose for which it will be tested, you can save valuable time and money relative to laboratory testing and waste management fees.

Representative Sampling: Requirements and Guidance

Where waste analyses are needed, it may be necessary to obtain certified data from a qualified, independent laboratory. As with all testing, there are two important caveats: (1) be certain that all analyses are performed on representative samples and (2) ensure that all samples are taken, handled and analyzed in accordance with state and federal standards. These two points can not be over emphasized. The data obtained from even the best laboratories are only as good as the sample submitted for analysis. There have been all too many cases where companies have spent hundreds and even thousands of dollars on laboratory analyses only to have the results discredited due to improper sample handling or storage, insufficient documentation for chain-of-custody or failure to obtain truly representative samples.

General Reference Materials
For a general overview and introduction to environmental sampling strategies, the reader is referred to Keith, L.H., *Environmental Sampling: A Summary* in Environmental Science and Technology, Vol. 24, No. 5, 1990.

This article discusses planning and sampling protocols, basic sampling approaches (random, systematic, and judgemental), migration of pollutants, the selection of sampling devices, safety planning, quality assurance, sampling plans for water (surface, precipitation, and groundwater), air, soils, solids, liquids, sludges and biological matrices. The environmental sampling publications edited/written by Keith, listed below, provide a more in-depth discussion of the topic:

- *Principles of Environmental Sampling,* Keith, L.H., Ed., American Chemical Society, 1155 16th St., N.W., Washington, D.C. 20036. 1988.

- *Principles of Environmental Sampling: Electronic Edition,* Keith, L.H., Ed., American Chemical Society, 1155 16th St., N.W., Washington, D.C. 20036. 1990.

- *Practical Guide for Environmental Sampling and Analysis,* Keith, L.H., Lewis Publishers, 2000 Corporate Blvd., N.W., Boca Raton, FL 33431. 1990.

Federal Standards and Guidance
As part of the federal standard for hazardous waste identification, the U.S. Environmental Protection Agency requires generators to obtain representative samples of wastes (for non-listed wastes) prior to testing for the characteristic of ignitability, corrosivity, reactivity or toxicity — 40 CFR Part 261 §§ 261.21-24, refer to Table 14. The specific language contained in 40 CFR § 261.23, for example, states that "a solid waste exhibits the characteristic of reactivity if a *representative sample* (emphasis added) of the waste has any of the following properties:

(1) It is normally unstable and readily undergoes violent change without detonating.

(2) It reacts violently with water.

(3) It forms potentially explosive mixtures with water.

(4) When mixed with water, it generates toxic gases, vapors or fumes in a quantity sufficient to present a danger to human health or the environment.

(5) It is a cyanide or sulfide bearing waste which, when exposed to pH conditions between 2 and 12.5, can generate toxic gases, vapors or fumes in a quantity sufficient to present a danger to human health or the environment.

(6) It is capable of detonation or explosive reaction if it is subjected to a strong initiating source or if heated under confinement.

(7) It is readily capable of detonation or explosive decomposition or reaction at standard temperature and pressure.

(8) It is a forbidden explosive as defined in 49 CFR §173.51, or a Class A explosive as defined in 49 CFR §173.53 or a Class B explosive as defined in 49 CFR §173.88.

This regulatory logic, therefore, dictates that a solid waste can not be a characteristic hazardous waste (other than based on generator knowledge) until and unless a "representative sample" of the waste tests positive for the characteristic in question. If a representative sample is not obtained, then the analysis can not be used in support of a legally or scientifically defensible waste determination.

For the purposes of guiding test procedures and sample collection efforts in support of hazardous waste determinations/RCRA compliance initiatives, the U.S. EPA has published 40 CFR Part 261 Appendix I *Representative Sampling Methods* and SW-846 *Test Methods for Evaluating Solid Waste, Physical/Chemical Methods* (Table 14).

SW-846 Chapter Nine "Sampling Plan" Overview

SW-846, Test Methods for Evaluating Solid Waste, is comprised of 2 volumes containing 13 chapters. Volume I contains 8 of the 13 chapters and is subdivided into two parts: Part I — Methods for Analytes and Properties, and Part II — Characteristics (i.e., regulatory definitions and laboratory methods for waste determinations). In Volume II, Chapter 9, guidance for the design and development of solid and hazardous waste sampling plans is presented. The remaining chapters of Vol. II discuss ground water monitoring, land treatment monitoring and incineration.

Chapter 9 provides detailed guidance on the design of a scientifically defensible solid and hazardous waste sampling plan. In the opening paragraph of this 79 page chapter, EPA describes the importance of obtaining representative samples:

> "The initial — and perhaps most critical — element in a program designed to evaluate the physical and chemical properties of a solid waste is the plan for sampling waste. It is understandable that analytical studies, with their sophisticated instrumentation and high cost, are often perceived as the dominant element in a waste characterization program. Yet, despite that sophistication and high cost, analytical data generated by a scientifically defective sampling plan have limited utility, particularly in the case of regulatory proceedings."[8]

Table 14
U.S. Environmental Protection Agency Representative
Sampling Requirements, Planning and Methods

40 CFR PART 261 SUBPART C — CHARACTERISTICS OF HAZARDOUS
WASTE: REPRESENTATIVE SAMPLING REQUIREMENTS
§261.20 GENERAL
§261.21 CHARACTERISTIC OF IGNITABILITY
§261.22 CHARACTERISTIC OF CORROSIVITY
§261.23 CHARACTERISTIC OF REACTIVITY
§261.24 TOXICITY CHARACTERISTIC
PART 261 APPENDIX I — REPRESENTATIVE SAMPLING METHODS
EXTREMELY VISCOUS LIQUIDS
FLY ASH-LIKE MATERIAL
CONTAINERIZED LIQUID WASTES
LIQUID WASTE IN PITS, PONDS, LAGOONS AND SIMILAR
RESERVOIRS

SW 846 — CHAPTER NINE: SAMPLING PLAN [a]
DESIGN AND DEVELOPMENT
DEVELOPMENT OF APPROPRIATE SAMPLING PLANS
Regulatory and Scientific Objectives
Fundamental Statistical Concepts
Basic Sampling Strategies
• Simple Random Sampling
• Stratified Random Sampling
• Systematic Random Sampling
Special Considerations
• Composite Sampling
• Subsampling
IMPLEMENTATION
DEFINITION OF OBJECTIVES
SAMPLING PLAN CONSIDERATIONS
Statistics
Waste
Site
Equipment
Quality Assurance and Quality Control
Health and Safety
Chain-of-Custody
SAMPLE PLAN IMPLEMENTATION
Containers, Tanks, Waste Piles, Landfills and Lagoons
SAMPLE COMPOSITING

[a] Source: U.S. Environmental Protection Agency, *Test Methods for Evaluating Solid Waste, Physical/Chemical Methods*, Office of Solid Waste and Emergency Response, Publication SW 846.

As shown in Table 14, Chapter 9 is divided into two major sections: (1) Design and Development, and (2) Implementation. The design and development section presents EPA guidance on regulatory and scientific objectives, statistical concepts and basic sampling strategies. Random sampling (simple, stratified and systematic), composite sampling and subsampling (i.e., replicate sampling) methods are described in detail. GENERATORS SHOULD NOTE that at the very outset of this chapter, EPA qualifies the information presented by stating that the "burden of responsibility for developing a technically sound sampling plan" rests with the generator and it is, therefore, in the generator's best interest to "seek competent advice" before developing and implementing such a plan.[9]

The implementation section of Chapter 9 consists of four subsections: definition of objectives, sampling plan considerations, sample plan implementation and sample compositing. This section provides basic guidance on how to write and implement a sampling plan in consideration of waste type/physical properties and method of waste storage or containerization. The description and use of waste sampling equipment, both specialized and nonspecialized, is reviewed; as shown in Table 15, EPA has developed a matrix that matches sampling equipment to waste type and method of storage. Quality assurance/quality control techniques, including the use of blanks, duplicates and spikes, are discussed. Health and safety, as well as chain-of-custody protocols are described.

The complete text of U.S. EPA guidance on representative sampling procedures and accepted laboratory test methods can be found in *Test Methods for Evaluating Solid Waste, Physical/Chemical Methods*, SW-846.[10] For the readers' convenience, key text from Chapter 9 —Sampling Plan— has been reprinted in Appendix D. The complete 2-volume set can be purchased from the U.S. Department of Commerce, National Technical Information Service, Springfield, VA 22161 (703/487-4650). The manual may also be available for review at your state environmental protection agency, commercial hazardous waste testing laboratory, university or U.S. EPA regional library.

TSDF Representative Sampling Requirements and Waste Tracking

In addition to the federal regulations/guidance cited above, generators must also comply with TSDF requirements for obtaining and submitting representative waste samples prior to material shipment. In order to ensure that samples are truly representative, generators are typically required to sign a sample certification statement similar to the one presented below.

The certification statement typically appears on a waste tracking label, which the generator applies to the sample container, or is incorporated into the overall generator certification statement that appears on the TSDF supplied waste data sheet (see Section on Generator Waste Data Sheets, Commercial Applications on p. 53).

Generator Certification Statement

I certify that the sample submitted herewith was collected in accordance with methods cited in "Test Methods for Evaluating Solid Waste, Physical/Chemical Methods, SW-846" and is representative of the waste described in the attached data sheet.

_____   _____

Signature   Waste Tracking No.

_____

Printed/Typed Name and Title

A FEW WORDS ON WASTE CHARACTERIZATION

Waste characterization remains one of the most challenging elements of the entire RCRA regulatory program. In addition to the expansive federal criteria, many states have established their own characteristics and/or listings. As a result, generators must not only learn and keep abreast of changes in the federal identification system, but must also develop a working knowledge of the classification systems in use in their state and in the state to which they ship. When offered for transportation, each hazardous waste must also be classified in accordance with the U.S. Department of Transportation's Hazardous Materials Regulations. Figure 18 shows the interrelationship, discussed below, between waste characterization and hazardous material classification programs.

HAZARDOUS WASTE UNIVERSE

Of the total universe of solid waste generated nationally, only a fraction is regulated as hazardous; according to EPA estimates,[12] of the 6 billion tons of industrial, agricultural, commercial and domestic waste generated annually, about 250 million tons are "hazardous" as defined by RCRA regulations. In some states, the universe of regulated hazardous waste includes all federally regulated hazardous waste (shown as "RCRA Hazardous Waste" in Figure 18) plus other waste materials which meet additional state criteria. Since all RCRA hazardous wastes are also regulated by the U.S. Department of Transportation (DOT) as hazardous material, each container of waste must be classified, packaged, marked and labeled according to DOT requirements when offered for transportation. Finally, the diagram shows three additional conditions that can exist: (1) some solid wastes can be regulated by a state as hazardous, but not by EPA or DOT, (2) certain state-regulated hazardous wastes are also DOT hazardous materials, but not

Table 15. Sampling Equipment Recommendation by Waste Type. [a]

	Waste Location or Container								
Waste type	Drum	Sacks and bags	Open-bed truck	Closed-bed truck	Storage tanks or bins	Waste piles	Ponds, lagoons, & pits	Conveyor belt	Pipe
Free-flowing liquids and slurries	Coliwasa	N/A	N/A	Coliwasa	Weighted bottle	N/A	Dipper	N/A	Dipper
Sludges	Trier	N/A	Trier	Trier	Trier	b	b		
Moist powders or granules	Trier	Trier	Trier	Trier	Trier	Trier	Trier	Shovel	Dipper
Dry powders or granules	Thief	Thief	Thief	Thief	b	Thief	Thief	Shovel	Dipper
Sand or packed powders and granules	Auger	Auger	Auger	Auger	Thief	Thief	b	Dipper	Dipper
Large-grained solids	Large trier	Large trier	Large trier	Large trier	Large trier	Large trier	Large trier	Trier	Dipper

[a] Source: U.S. Environmental Protection Agency, *Test Methods for Evaluating Solid Waste, Physical/Chemical Methods.*

[b] This type of sampling situation can present significant logistical sampling problems, and sampling equipment must be specifically selected or designed based on site and waste conditions. No general statement about appropriate sampling equipment can be made.[11]

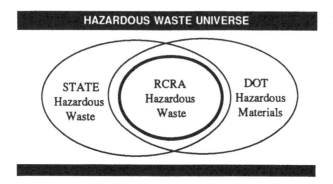

FIGURE 18. Relationship between RCRA and state hazardous waste classification programs and the U.S. Department of Transportation's hazardous materials regulations.

RCRA hazardous waste, and (3) some solid wastes, though not regulated as RCRA or state hazardous waste, can be regulated by DOT as hazardous material when offered for transportation, e.g., discarded asbestos.

EVALUATION AND CLASSIFICATION OF WASTES

In general, wastes are evaluated and characterized according to specific criteria established by the state and federal government. At the federal level, generators determine whether their wastes are hazardous by reviewing standard listings and by comparing waste properties with established criteria. Regulatory standards are then applied according to the type of waste generated and the proposed method for recycling, treatment or disposal. The waste evaluation procedure established under the authority of RCRA, as described above, consists of three principal steps:

Section 262.11 of Title 40 Code of Federal Regulations (CFR) requires a generator of a "solid waste" to: [13]

1. "Determine if [the] waste is excluded.
2. If it is not excluded, [the generator] must determine if the waste is listed as a hazardous waste.
3. If the waste is not excluded and not listed, then [the generator] must evaluate the waste in terms of the four Hazardous Characteristics, unless [the generator] can properly evaluate the waste based upon his/her own knowledge of the waste (e.g., corrosivity testing may not be required if the generator has a long history of running the waste through steel pipes without any evidence of corrosion)."

In a practical sense, applying these steps to some types of waste is not always easy. It is not unusual, for example, for a waste to exhibit more than one hazardous characteristic (e.g., spent chromic acid would be classified as both reactive and corrosive) or to be excluded from some regulations and not others. In other cases, sufficient data may not exist or information on the potential hazard or history of the waste may be obscured or even lost. And finally, there are those cases where the situation is so complex that it forces the generator to seek legal advice or other outside assistance. Whatever your situation may be, always keep in mind that it is the legal responsibility of each and every generator to properly classify and manage its own hazardous waste.

Material Composition and Waste Classification

When you stop to think about it, everything you do to a waste after it has been generated is determined, for the most part, by its material composition. Failure to understand the properties and chemical nature of your waste can have a profound and lasting impact on your program. Improper characterization could trigger a "domino effect," setting off a chain reaction of potentially hazardous or illegal events. A hazardous waste that has been erroneously classified, for example, may be improperly, illegally and/or unsafely labeled, packaged, handled, stored, transported and treated or disposed of.

Because waste characterization is such an important element of your program, you should make every effort to obtain a complete and accurate profile of each waste generated at your facility. Take the time to read and understand the regulations. Never leave the waste characterization task entirely up to a vendor or someone who is not familiar with your waste generating processes, operations or activities. A mistake, made by someone other than yourself, may come to bear on you, the generator. Always retain control and acquire the necessary skills, technical competence or training where needed. Also, be sure to thoroughly investigate all solid wastes for potential hazard. Some wastes, though seemingly harmless, prove to be hazardous waste when subjected to laboratory analysis. Here are two examples:

- In one report,[14] "laboratory data resulting from the analysis of crushed glass from a fluorescent light bulb revealed a mercury level of 0.98 mg/l [extractable metal] which is well above the 0.2 mg/l which is the concentration that designates a D009 hazardous waste." In this report, the author explains that it is standard practice for most companies to discard old fluorescent bulbs in the regular trash "without ever considering the fact that mercury is an integral part of the chemistry of the bulb."
  [NOTE: A more recent study conducted by the U.S. Environmental Protection Agency Office of Air Quality's Control Technology Center

(CTC)[15] further reports that a "typical four-foot fluorescent lamp contains about 41 mg (total metal) of mercury." Authors of the EPA CTC study state that in 1991 alone, approximately 500 million fluorescent lamps were manufactured and that production in subsequent years would likely increase as a direct result of EPA's Green Lights program. From a management perspective, EPA's Office of Air Quality is concerned with the total amount of mercury emitted to the atmosphere when discarded. Researchers note that the "total amount of mercury emitted to the atmosphere depends on how a spent lamp is handled. If lamps are broken in a garbage truck on-route to a landfill, most of the mercury can find its way into the atmosphere. However, if spent lamps are packed in corrugated containers, delivered to a landfill in enclosed vans or trailers, and placed in the fill with minimal breakage, practically all the mercury would be retained in the landfill."[16] For information on how to recover and recycle mercury from fluorescent lamps, the reader is referred to the original study. Copies of the report may be obtained from the U.S. Environmental Protection Agency, Office of Air Quality Planning and Standards, Control Technology Center, (MD-13) Research Triangle Park, NC 27711.]

• In another study, one company set out to replace old, leaking rain gutters from a small building with newer ones. When the engineer in charge of the project sent a sample of the gutter out for laboratory analysis, results showed lead levels in the liquid extract to be in excess of 5 mg/l. Ultimately, the lead-coated copper gutters had to be managed as hazardous waste.

Sources of Information

In addition to Title 40 of the Code of Federal Regulations (CFR) and related daily federal register publications, there are a number of references that can be of help throughout your waste characterization and management efforts. The publications listed in Table 16 contain chapters that analyze the RCRA hazardous waste identification system. One book, the *Hazardous Waste Identification and Classification Manual,* reviewed by C.A. Evans,[17] has been described as a "comprehensive guide providing lengthy and complex information in an easy to understand format." In her review, Evans says that thirty-five, real-life case studies focusing on the hazardous waste identification process are contained in one chapter, while another chapter gives step-by-step procedures for classifying hazardous waste for transportation. The book is written by Travis P. Wagner (240 pages) and is published by Van Nostrand Reinhold.

Additional sources of related information can be found in Guidance Documents published by the U.S. Environmental Protection Agency.

Table 16
Selected Hazardous Waste Reference Materials

Arbuckle, J.G. et al., *Environmental Law Handbook,* Government Institutes, Inc., 966 Hungerford Drive #24, Rockville, Maryland 20850. Eleventh Edition. 1991.

Fournier, S., *Handbook of Hazardous Waste Regulation, Volume I: Comprehensive Introduction to RCRA Compliance,* Business & Legal Reports, Inc., 64 Wall Street, Madison, CT 06443. September 1992.

Hall, R.M. et al., *RCRA Hazardous Wastes Handbook,* Government Institutes, Inc., 966 Hungerford Drive #24, Rockville, Maryland 20850. Nineth Edition. 1991.

J.J. Keller & Associates, Inc., *Hazardous Waste Management Guide,* 145 West Wisconsin Ave., P.O. Box 368, Neenah, WI 54956-0368. 1992.

Stever, D.W., *Law of Chemical Regulation and Hazardous Waste,* Clark Boardman Company, Ltd., 155 Pfingsten Road, Deerfield, IL 60015-9917. 1992.

Wagner, T.P., *Hazardous Waste Identification and Classification Manual,* Van Nostrand Reinhold, 135 West 50th Street, New York, New York 10020. 1990.

Examples include the *Waste Analysis Plan Guidance Manual* (EPA/530-SW-84-012, October 1984), *A Guide for Estimating the Incompatibility of Selected Hazardous Waste Based on Binary Chemical Mixtures* (ASTM designation #P-168), *Compatibility of Wastes in Hazardous Waste Management Facilities* (RCRA Docket), and *Petitions to Delist Hazardous Waste: A Guidance Manual* (EPA/530-SW-85-003, NTIS #PB 85-194-488). Guidance documents have been written on a wide range of solid and hazardous waste management topics and are available from three principal sources:

- National Technical Information Service, 5285 Port Royal Road, Springfield, VA 22161. Phone: (703) 487-4650
- U.S. Environmental Protection Agency, 401 M Street, Washington, D.C. 20460. RCRA Hotline: 1 (800) 429-9346
- Superintendent of Documents, Government Printing Office, Washington, D.C. 20402-9325. Phone: (202) 783-3238

As a final note, you should strive to develop a sound understanding of the federal register system and the Code of Federal Regulations. An excellent guide for users of the Federal Register is *THE FEDERAL REG-*

ISTER: What It Is And How To Use It. This guide is published by the Office of the Federal Register, National Archives and Records Service, General Services Administration, Washington, DC 20408.

DETAILED RECORDKEEPING GUIDANCE

Once complete, waste data sheets will prove to be a valuable resource for environmental managers. As described earlier, recorded information and data can be used, for example, to warn workers of the risks involved with improper waste management.

Data sheets should be reviewed in detail with all personnel who are or may be required to handle hazardous waste. At the time of each off site shipment, managers may wish to review personal protection equipment requirements, safety policies and emergency response procedures with material handlers and transportation personnel. Where acceptable to regulatory authorities, waste data sheets may also be incorporated into your facility's Contingency Plan, distributed to local fire departments and hospitals, or kept near waste storage areas.

In general, the more complete and reliable your data set, the better prepared you will be to meet waste characterization, material tracking and other technical demands.

DATA SHEET DESIGN

Figure 13 provides one example of a simple waste data sheet that was created using an Aldus Pagemaker® program. As with all sample record-keeping forms presented in this book, generator waste data sheets can be designed in-house using a computer graphics software program (see Chapter 2, footnote page 17) or sketched by hand and then submitted to a local print shop for professional typesetting and duplication.

SAMPLE WASTE DATA SHEET:
COMMENTS AND GUIDANCE

Located in the upper right-hand corner of the waste data sheet (Figure 13) is a line for recording the waste ID number or source identifier. The main body of the sample data sheet is divided into eight (8) sections. Each section has been carefully designed to allow for the collection of the most important and most frequently used information and data.

One data sheet should be completed for each waste generated. For the most part, data entry blocks on the sample form (Figures 13 and 14) require little or no introduction. Unfamiliar terms can be found in basic science texts or chemical dictionaries. When reviewing the sample data sheet presented on pages 56 and 57, the reader should keep in mind that it was designed as a "universal" form to be used as a guide by generators of all types. If this format is adopted by you, therefore, there may be data entry blocks that are not applicable to a particular process or waste stream. In such cases, data blocks may be skipped over or appropriately noted — e.g., "not analyzed for" or "not applicable." All other pertinent data and information should be recorded in the additional space provided. The following line-by-line guidance is keyed to the generic data sheet format shown in Figures 13 and 14, and is intended to assist you in the organization of your data.

Section I — General Waste Description

Process and waste descriptions should be consistent with the ones assigned in Chapter 2. Rate of waste generation can be expressed as a numerical range or average value per unit time. Maximum amounts and annual totals should also be recorded here. Indicate whether the pattern of waste generation is continuous, discrete or subject to seasonal fluctuation.

Section II — Chemical Composition

Provide the chemical name, formula and Chemical Abstract Service (CAS) number for each waste constituent. The Chemical Abstract Service is a division of the American Chemical Society, a not-for-profit scientific and educational institution chartered by the U.S. Congress and headquartered in Washington, DC.[18] CAS abstracts and indexes chemical literature from all over the world in Chemical Abstracts. "CAS Numbers" are used to identify specific chemicals or mixtures.[19] The CAS number may be used in literature searches and government reports and is sometimes requested by off site TSDFs as part of their waste acceptance procedure.

Chemical concentrations may be recorded as a range with an average value, (10 - 30%) 25, for example. Record concentrations in percentages, parts per million (ppm) or other units as appropriate.

Sections III, IV and V — Physical/Reactive Properties, Metals and Other Components

All data that characterize a waste's physical, chemical, and reactive properties should be recorded in these sections. Provide actual test data where available. Report units of measurement, analytical method and test conditions where applicable; for example, Flash point: 132°F, Tag Closed Tester (ASTM D56) or Metals, Lead: 5 mg/l TCLP. Incompatible materials, conditions to avoid and hazardous decomposition products may be recorded in the additional space provided in Section IV. Guidance for pre-

---

ERG93   **GUIDE 53**

POTENTIAL HAZARDS

HEALTH HAZARDS

Poisonous if swallowed.

Inhalation of dust or mist may be poisonous.

Fire may produce irritating or poisonous gases.

Runoff from fire control or dilution water may cause pollution.

FIRE OR EXPLOSION

Some of these materials may burn, but none of them ignites readily.

EMERGENCY ACTION

Keep unnecessary people away; isolate hazard area and deny entry.

Stay upwind; keep out of low areas.

Positive pressure self-contained breathing apparatus (SCBA) and structural fire fighters' protective clothing will provide limited protection.

CALL Emergency Response Telephone Number on Shipping Paper first.
If Shipping Paper not available or no answer, CALL CHEMTREC AT 1-800-424-9300 FOR EMERGENCY ASSISTANCE.

If water pollution occurs, notify the appropriate authorities.

FIRE

Small Fires: Dry chemical, CO2, water spray or regular foam.

Large Fires: Water spray, fog or regular foam. Move container from fire area if you can do it without risk.

SPILL OR LEAK

Do not touch or walk through spilled material; stop leak if you can do it without risk.

Small Spills: Take up with sand or other noncombustible absorbent material and place into containers for later disposal.

Small Dry Spills: With clean shovel place material into clean, dry container and cover loosely; move containers from spill area.

Large Spills: Dike far ahead of liquid spill for later disposal.

FIRST AID

Move victim to fresh air; call emergency medical care.

Remove and isolate contaminated clothing and shoes at the site.

In case of contact with material, immediately flush skin or eyes with running water for at least 15 minutes.

Figure 19. Emergency Response Guide for nickel cyanide. Guidebook for initial response to hazardous materials incidents, U.S. DOT RSPA P 5800.6, for sale by the U.S. Government Printing Office, Superintendent of Documents, Mail Stop: SSOP, Washington, D.C. 20402-9328.

dicting the reactive properties of wastes may be found in the publication entitled *Design and Development of a Hazardous Waste Reactivity Testing Protocol*, listed in Table 10. This manual presents a test scheme "to classify wastes into reactivity groups and thus predict compatibility."

Section VI — Health Hazard Data and Personal Protection

Information needed to complete this section may come from a variety of sources including government and private publications (e.g., IRIS and ATSDR Toxicological Profiles), computer data bases, Material Safety Data Sheets and company-commissioned bioassays. Describe principal routes of entry and safe handling procedures. List required personal protective equipment such as gloves, boots, safety glasses, face shield, protective suit, air-purifying or air-supplying respirators.

Section VII — Spill, Fire and Emergency Response

Note fire and explosion hazard data, fire fighting procedures, and spill and emergency response procedures in this section. List the names, addresses and phone numbers (office and home) of all persons qualified to act as emergency coordinator[20] for the site in this section or in the header of each data sheet.

Data collected under the U.S. DOT's Emergency Response Information Communication Standard (refer to the "Communicating Waste Material Hazards" section of this chapter) may be recorded here. Where more specific information is not available, spill, fire and emergency response information can be found in the *DOT Emergency Response Guidebook*, a guidebook for initial response to hazardous materials incidents, RSPA P 5800.6. This guidebook is published by the U.S. Department of Transportation, Research and Special Programs Administration, Office of Hazardous Materials Training and Initiatives (DHM-50), Washington, D.C. 20590-0001. A sample emergency response guidebook page for nickel cyanide is presented in Figure 19.

Section VIII — Regulatory Information

This section is provided for recording the results of your waste characterization and hazardous material classification efforts. The data obtained in Sections II, III, IV and V will assist you in completing this section. Enter the U.S. Department of Transportation and U.S. Environmental Protection Agency waste material descriptions, regulatory classifications and land ban information here. Other information that may be recorded in this section includes:

- Labeling and special marking requirements
- Method of shipment, treatment and disposal
- Placarding requirements
- Hazardous substance reportable quantities

You should also note the basis for your waste characterization in the space provided. Assistance may be available from state or federal regulatory agencies/technical assistance programs or commercial laboratories.

Record all other important information in the space provided under NOTES; for example, OSHA-regulated substances that are present in the waste, current or pending treatment, or disposal options and approvals. You may also want to specify drum labeling and marking requirements here. The data sheet could then be issued to the person(s) responsible for preparing shipments.

END NOTE

Very often, hazardous waste exhibits properties similar to the raw materials used in the generating process or operation. In the case of obsolete inventory, the properties of the discarded product would be exactly the same as those indicated on the Material Safety Data Sheet or found in the literature. A list of reference materials and bibliographic data bases on chemical properties of raw materials is provided in Table 10 to help you characterize these types of wastes.

In other cases, waste properties can not be accurately predicted and representative samples of the waste must be subjected to instrumental analysis or other forms of laboratory testing. In either case, whenever supportive data (e.g., MSDSs or certificates of laboratory analysis) are used or relied upon for waste characterization or for any other purpose, you should (1) annotate the waste data sheet accordingly, and (2) where practical, record the waste identification number or source identifier on copies of each available document and bind each document copy with its appropriate data sheet.

REFERENCES

1. **Environmental Systems Company (ENSCO),** Waste Material Data Sheet: Instructions, American Oil Road, El Dorado, AR 71730.
2. **40 Code of Federal Regulations** §262.11.
3. **U.S. Department of Labor,** *Hazard Communication: A Compliance Kit,* OSHA Publication No. 3104. 1988.

REFERENCES (Continued)

4. U.S. Environmental Protection Agency, *Integrated Risk Information System*, Office of Research and Development, Environmental Criteria and Assessment Office, Cincinnati, Ohio 45268.

5. Amdur, M.O. et al., *Casarett and Doull's Toxicology, The Basic Science of Poisons,* McGraw-Hill, Inc., 1221 Avenue of the Americas, New York, New York 10020. Fourth Edition. 1991.

6. Morse, K., performed RTECS data base search for benzene. Mr. Morse is associated with the Reference Unit, Public Services Department, University of Rhode Island Library, Kingston, Rhode Island 02881.

7. op. cit. Amdur, M.O. et al.

8. U.S. Environmental Protection Agency,*Test Methods for Evaluating Solid Waste, Physical/Chemical Methods,* Third Edition, Office of Solid Waste and Emergency Response, Publication SW-846 (November 1986/revisions December 1987, Washington, D.C. 20460.

    Available from the U.S. Department of Commerce, National Technical Information Service, Springfield, VA 22161, 703/487-4650 (document number 955-001-00000-1).

9. ibid.

10. ibid.

11. ibid.

12. U.S. Environmental Protection Agency, *Solving the Hazardous Waste Problem: EPA's RCRA Program,* Office of Solid Waste, Washington, DC 20460. November 1986.

13. op. cit. U.S. Environmental Protection Agency, SW-846.

14. Strand, S.A., *Waste Management Concerns for the 90s, Manufacturers' Mart Publications,* P.O. Box 723, Fairfield, CT 06430. September 1991.

15. U.S. Environmental Protection Agency, *Control Technology Center News,* Office of Air Quality Planning and Standards, Research Triangle Park, NC 27711. April 1994.

16. ibid.

17. Evans, C.A., *Hazardous Waste: Can I See Your I.D., Please?,* Environmental Waste Management Magazine, 243 West Main St., Kutztown, PA 19530. November 1990.

    The book *Hazardous Waste Identification and Classification Manual,* on which the above review article is based, is written by T.P. Wagner and published by Van Nostrand Reinhold, 135 West 50th Street, New York, New York 10020. 1990.

18. Chemical Abstract Service, *1990 Current-Awareness Catalog,* 2540 Olentangy River Road, P.O. Box 3012, Columbus, Ohio 43210.

19. op. cit. U.S. Department of Labor.

20. 40 Code of Federal Regulations §265.52(d).

Chapter 5

OVERVIEW OF SHIPMENT AND COST
DATA REQUIREMENTS FOR WASTE TRACKING

INTRODUCTION

In Chapters 2 through 4, the groundwork was laid for waste material tracking by point of generation. The identity, origin and properties of each waste were clearly defined and systematically recorded. In this chapter, your hazardous waste tracking program will be expanded by broadening the scope of your data collection efforts to include *off site* activity tracking and "fully-loaded" waste management cost accounting. The sections which follow outline the general categories of data that will be needed to complete the final step (presented in Chapter 6) of your point of generation tracking and cost accounting program.

SHIPMENT AND COST DATA REQUIREMENTS

Over time, generators orchestrate the flow of countless sheets of paper including manifests, bills of lading, certificates of disposal/destruction, purchase orders, invoices, accounts payable records and related tracking and cost accounting documents. Many of these documents, however, are filed separately or reviewed jointly only in response to a specific request or to meet an immediate need. Moreover, waste management data contained in these records are not always organized in a way that facilitates unit tracking and full cost accounting. As a result, administrative "firefighting" often occurs where important details are obscured or even lost in the paper chase.

In the following sections, selected waste management and cost data are identified, gathered and then organized into a simple, yet effective, spreadsheet accounting system. Before the details of this approach are discussed in Chapter 6, however, it is helpful to review the types of data and information that will be needed. Data and information needed to complete the final step of your tracking program can be grouped into the following three categories:

- Manifest and Point of Generation Data
- TSDF Final Disposition Report
- Complete Management Cost Data

93

Figure 20. Tally sheet used at the time of shipment to sum wastes by point of generation.

A brief overview of each of the above categories is provided below. In general, most of the data that you will need to track wastes off site and relate quantities and costs back to the point of generation already exist and are readily available to you in the form of manifests, reports and accounts payable records. In some cases, however, you may need to expand the scope of your data collection efforts to acquire new or additional information.

MANIFEST AND POINT OF GENERATION DATA

Much of the information needed for "off site" waste tracking will come from the Uniform Hazardous Waste Manifest. This information includes: *Shipment Date, Manifest Document No., U.S. EPA/State Hazardous Waste Number(s), Container, Amount of Waste, Transporter and Designated Facility.* To facilitate recordkeeping efforts, you should organize your manifests chronologically by shipment date. The use of manifest data in unit tracking by point of generation is described in Chapter 6.

Depending upon the nature of the shipment, you may also need to tally each unique waste by point of generation. Very often, chemically similar wastes generated at several unique locations are added together and recorded as one entry on the manifest form; see Case Study, Chapter 2, page 12. When this happens, a waste's history of origin can be lost if a separate tally sheet is not kept at the time of shipment. The tally sheet is simply a notepad or piece of paper on which each unit of waste is listed separately by point of generation. As containers or bulk wastes are loaded onto the transport vehicle, a mark representing each container (or the total volume/weight of each waste) is recorded next to the source identifier as shown in Figure 20. Once the shipment has been made, this information is then transcribed into a daily journal as described in Chapter 6.

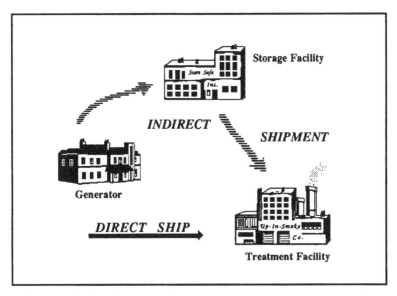

Figure 21. Direct and indirect generator shipments to a commercial treatment facility.

TSDF FINAL DISPOSITION REPORT

When shipping hazardous waste off site for treatment or disposal, generators can make one of two choices. They can either (1) arrange for *direct shipment* of their waste to a permitted treatment/disposal facility, or (2) employ the services of a storage/transfer facility (*indirect shipment*), as shown in Figure 21 above.

Direct Shipments

When a generator initiates a *direct shipment* to a commercial treatment or disposal facility, the ultimate disposition of the waste will rarely be in question. Since the generator communicates directly with the treatment/ disposal facility, the details concerning unit costs and waste management methods are generally well understood, or at least directly accessible. Direct shipments, though somewhat easier to track, are not always possible to make due to small quantities, travel distances or other factors that make them economically impractical.

Commercial Incineration and Land Disposal Facilities
Tables 17 and 18 list selected RCRA treatment (i.e., incineration) and disposal facilities, respectively, that were reported in the U.S. EPA's National Oversight Database and related EPA publications. The active, commercial facilities listed in these tables receive solid and/or hazardous waste shipments from generators (direct shipments) and from qualified storage/

Table 17. List of Commercial Facilities with RCRA Incinerators That May Be Operational before 1993. [a]

U.S. EPA Region	EPA ID Number	Facility Name/Location	Type of Kiln (No. of Units: MBtu/hr)	Waste Feed [b] Capability (Status)
REGION I	None Listed [c]			
REGION II New York	NYD000632372	BDT, Inc. 4255 Research Parkway Clarence, NY 14031 716/759-6901	Fixed hearth with secondary combustion chamber (1:2)	Solids
New Jersey	NJD053288239	Rollins Environmental Services (NJ) RTE 322 & 295 Bridgeport, New Jersey 08014 609/467-3100	Rotary kiln w/ liquid injection (1:35) Liquid injection (1:75) Rotary kiln w/ liquid injection (1:36)	Liquids and solids Liquids only Liquids, solids and sludges
	NJD002385730	DuPont E I De Nemours & Co. Deepwater Chambers Works Route 130 Deepwater, NJ 08023 609/299-5000	Rotary kiln w/ liquid injection	DNA
REGION III	None Listed [c]			
REGION IV Kentucky	KYD088438874	LWD, Inc. Clay, Kentucky	CBI	CBI (DI)

Continued

Table 17. (Continued)

U.S. EPA Region	EPA ID Number	Facility Name/Location	Type of Kiln (No. of Units: MBtu/hr)	Waste Feed Capability (Status)
	KYD08843817	LWD, Inc. Calvert City KY Highway 1523 Calvert City, Kentucky 42029 502/395-8313	CBI	CBI
	KYD006396246	Olin Corporation Chemicals Brandenburg, Kentucky 502/422-2101	Liquid injection (2: 40 ea.)	Liquids only
North Carolina	NCD006871282	Caldwell Systems, Inc. MT Hermon Road Lenoir, North Carolina 28645	Liquid injection (1: 14)	Liquids (C)
South Carolina	SCD058754789	Groce Laboratories Greer, South Carolina 803/879-2020	Pyrolytic destructor (1: 1)[d] Deflagrating reactor (2: NA)[e]	Solids Liquids, solids and sludges
	SCD981467616	Laidlaw Env Svs (TOC) Inc. 301 Railroad Street Roebuck, South Carolina 29376 803/576-6821	Liquid injection (1: 50) Rotary kiln (1: 60)	Liquids only Solids and sludges only
	SCD944442333	Thermal Kem, Inc. (formerly Stablex) RTE 5 Vernesdale Road Rock Hill, SC 29731 803/324-5310	Fixed hearth (2: 42 ea.)	Both accept liquids, sludges, solids, and gases

Continued

Table 17. (Continued)

U.S. EPA Region	EPA ID Number	Facility Name/Location	Type of Kiln (No. of Units: MBtu/hr)	Waste Feed Capability (Status)
REGION V Illinois	ILD098642424	Chemical Waste Management, Inc. Incineration Division 7 Mobile Avenue Sauget, Illinois 62201 618/271-2804	Fixed hearth (3: 14, 16) Rotary kiln (1: 16)	Liquids, solids and sludges
	ILD000672121	CWM Chemical Services (formerly SCA) 11700 S. Stony Island Ave. Chicago, Illinois 60617 312/646-5700	Rotray kiln w/ liquid injection afterburner (1: 120)	Liquids, solids and sludges
Ohio	OHD048415665	Ross Incineration Services, Inc. 36790 Giles Road Grafton, Ohio 44044 216/748-2171	CBI	CBI
	OHD980569438	GSX Chem. Services of Ohio, Inc. 7415 Bessemer Ave. Cleveland, Ohio 44127 216/441-5628	Infrared w/ secondary combustion chamber (1: 30)	Solids and sludges (D)
REGION VI Arkansas	ARD069748192	Environ. Systems Co. Inc. (ENSCO) American Oil Road El Dorado, AR 71730 501/863-7173	Rotary kiln w/ liquid injection (3: 170, 36)	Liquids, solids and sludges

Continued

Table 17. (Continued)

U.S. EPA Region	EPA ID Number	Facility Name/Location	Type of Kiln (No. of Units: MBtu/hr)	Waste Feed Capability (Status)
Louisiana	LAD010395127	Rollins Environmental Services 13351 Scenic Highway Baton Rouge, LA 70807 504/778-1234	Rotary kiln w/ liquid injection (2: 131, 150)	Liquids, solids and sludges
	LAD008161234	Stauffer Chemical Co., Inc. Baton Rouge, Louisiana	Liquid injection, H2SO4 regenerating furnaces (2: 100, 180)	Liquids and sludges
Texas	TXD008099079	Stauffer Chemical Co., Inc. Houston, Texas 703/688-3495	Liquid injection (1: 205)	Liquids and sludges
	TXD055141378	Rollins Environmental Services 2027 Battleground Road Deer Park, Texas 77536 713/930-2300	Rotary kiln w/ liquid injection (1: 150); Rotary kiln w/ liquid injection (1: 100); Rotary reactor (2: 36, 34)	Liquids, solids sludges and gases
	TXD000838896	Chemical Waste Management, Inc. Highway 73 W. 3M. Port Arthur, Texas 77640 409/736-2821	Rotary kiln w/ liquid injection (1: 150)	Liquids, solids and sludges (P)
REGION VII Missouri	MOD073027609	Industrial Service Corp. (formerly Radium Petroleum Co.) 1633 S. Marsh B Kansas City, Missouri 64126 816/833-1919	Rotary kiln w/ liquid injection (1)	Liquids, solids and sludges (D)

Continued

Table 17. (Continued)

U.S. EPA Region	EPA ID Number	Facility Name/Location	Type of Kiln (No. of Units: MBtu/hr)	Waste Feed Capability (Status)
Kansas	KSD981506025	Aptus Environmental Services Coffeyville, Kansas 316/251-6380	Rotary kiln w/ liquid injection (1: 62)	Liquids and solids
REGION VIII	None Listed [c]			
REGION IX California	CAD042245001	Omega Chemical Corp. Whittier, California	Liquid injection (1: 10) Fluidized bed (1: 10)	Liquids (DI)
	CAD000094771	IT Corp. Martinez, California	Liquid injection (1: 18)	Liquids/solids (S, 1988)
REGION X	None Listed [c]			

CBI — Confidential business information
NA — Not available at this time
DNA — Data not available

[a] SOURCE: (1) U.S. Environmental Protection Agency, *Treatment, Storage and Disposal Facilities with Commercial Processes*, RCRIS National Oversight Database, Washington, D.C. 20460, August 1994, and (2) U.S. Environmental Protection Agency, *Commercial Treatment/Recovery Capacity Data Set* (PB90-259789), Washington, D.C. 20460, May 1990.

[b] Facilities operational unless specifically noted as — operating status delayed (D), delayed indefinitely (DI), planned (P), closed (C), or stopped accepting wastes (S).

[c] There are no commercial facilities with RCRA incinerators listed in the U.S. EPA RCRIS (1994) database for this region.

[d] A pyrolytic destructor is a thermal treatment method, but is not considered to be an incinerator (U.S. EPA, 1990).

[e] A deflagrating reactor is a thermal treatment unit, not an incinerator, that is used for the destruction of explosive wastes (U.S. EPA, 1990).

Table 18. Operating Commercial Land Disposal Facilities and RCRA Disposal Facilities with Commercial Processes. [a]

U.S. EPA Region	EPA ID Number	Facility Name	Facility Data Facility Address	Telephone No.	TSD [b] Type
REGION I	None Listed [c]				
REGION II					
New Jersey	NJD002385730	Dupont E I De Nemours & Co. Deepwater	Chamber Works Route 130 Deepwater, NJ 08023	609/299-5000	D T S01/2
New York	NYD080336241	CECOS International Inc.	56th St. & Pine Ave. Niagara Falls, NY 14302	716/731-3281	(C)
	NYD049836679	CWM Chemical Services Inc.	1550 Balmer Road Model City, NY 14107	716/285-8208	D80 S01/2 T01/2/4
Puerto Rico	PRD091018622	Protecion Tecnica Ecologica	Carr 385 KM 3 HM 5 Tall Aboa Penuelas, PR 00724	809/836-2058	D80 S
REGION III					
Maryland	MDD000731356	Hawkins Point Area 5	5501 Quarantine Street Baltimore, MD 21226	301/974-7295	D80
Pennsylvania	PAD004835146	Mill Service Inc. Yukon Plant	Rd. No. 1 Cemetary Lane Yukon, PA 15698	412/343-4906	D S

Continued

Table 18. (Continued)

		Facility Data			
U.S. EPA Region	EPA ID Number	Facility Name	Facility Address	Telephone No.	TSD Type
Pennsylvania (Cont'd)	PAD085690592	Republic Environmental Systems	2869 Sandstone Drive Hatfield, PA 19440	215/822-8996	D80 S01/2/3 T03
REGION IV Alabama	ALD000622464	Chemical Waste Management Inc.	Highway 17 N/MI Marker 163 Emelle, AL 35459	205/652-9721	D80 S01 T02/4
Georgia	GAD033582461	Alternate Energy Resources Inc.	2730 Walden Drive Augusta, GA 30904	404/738-1571	D S01
South Carolina	SCD070375985	Laidlaw Env. Svs. of SC Inc.	RTE 1 Box 255 Pinewood, SC 29125	803/452-5003	D80
	SCD070371885	Phibro-Tech Inc. (formerly CP Chemicals)	Hwy 15 S. Industrial Park Sumter, SC 29150	803/481-8528	D
REGION V Illinois	ILD980700728	Browning Ferris Ind. of ILL Inc.	9th Street & Green Bay Road Zion, IL 60099	708/746-5777	D

Continued

Table 18. (Continued)

U.S. EPA Region	EPA ID Number	Facility Name	Facility Address	Telephone No.	TSD Type
			Facility Data		
Illinois (Cont'd)	ILD010284243	CID Landfill (a CWM facility)	138th & Calumet Expy Calumet City, IL 60409	312/891-1500	D80
	ILD074411745	CWM Laraway RDF	RTE 1 Laraway Road Elwood, IL 60421	815/727-6148	D S
	ILD000805812	Peoria Disposal Co.	4349 Southport Road Peoria, IL 61615	309/688-0760	D80 S01/2 S03/4
Indiana	IND078911146	Chemical Waste Management of Indiana Inc.	4636 Adams CTR. Road Fort Wayne, IN 46806	219/447-5585	D80 S
	IND000780544	Four County Landfill	821 N. Michigan Rochester, IN 46975	219/563-8706	D80
	IND980503890	Heritage Environ. Serv. Inc. ILWD	RR No. 1 Roachdale, IN 46172	317/243-0811	D80
	IND093219012	Heritage Environmental Services	7901 W. Morris Street Indianapolis, IN 46231	317/243-0811	D80
	IND980503775	ILWD Landfill	RR. 3 Box 156 Columbus, IN 47201	317/243-0811	D
Michigan	MID048090633	Wayne Disposal Inc. Site #2	49350 N. I94 Service Dr. Belleville, MI 48111	313/697-7830	D80
Ohio	OHD980795264	Browning Ferris Ind. of Ohio Inc.	6201 Pleasant Valley Rd. E. Palestine, OH 44413	216/747-4433	D80

Continued

Table 18. (Continued)

U.S. EPA Region	EPA ID Number	Facility Data			
		Facility Name	Facility Address	Telephone No.	TSD Type
Ohio (Cont'd)	OHD087433744	CECOS Intl. Inc.	5092 Aber Road Williamsburg, OH 45176		C
	OHD020273819	Chemical Waste Management	3956 St. RTE 412 Vickery, OH 43464	419/547-7791	D80
	OHD000721415	Envirosafe Ser. Wynn Rd.	Cedar Pt. and Wynn Rd. Oregon, OH 43616	419/726-1521	D
	OHD045243706	Envirosafe Services Otter Creek Rd.	876 Otter Creek Rd. Oregon, OH 43616	419/726-1521	D80 S01/2 S03/4 T01/4
	OHD055522429	Evergreen Environmental Group Inc.	33 Industry Dr. Bedford, OH 44146	216/439-1257	D80 S
	OHD980793384	Reserve Environmental Services	4633 Middle Road Ashtabula, OH 44004	216/992-6143	D80 S
Wisconsin	WID098547854	Metro Recycling and Disposal Facility	10712 S. 124th St. Franklin, WI 53132	414/529-6180	D80
REGION VI Louisiana	LAD000618256	CECOS International Inc.	Hwy. 397 Willow Springs Rd. Westlake, LA 70669	318/527-6857	D99

Continued

Table 18. (Continued)

			Facility Data		
U.S. EPA Region	EPA ID Number	Facility Name	Facility Address	Telephone No.	TSD Type
Louisiana (Cont'd)	LAD000618298	CECOS International Livingston Facility	28422 Frost Road Livingston, LA 70754		C
	LAD000777201	Chemical Waste Management Inc.	Hwy 108 2 MW Hwy 27 Carlyses, LA 70663	318/583-2169	D80 S01/2 T01/4
Oklahoma	OKD065438376	US Pollution Lone Mountain	15M S. on H281 5M E. H15 & 1M N. Waynoka, OK 73860	405/528-8371	D80 S01/2 T04
Texas	TXD000761254	The GNI Group (Disposal Systems Inc.)	6901 Greenwood Rd. Corpus Christi, TX 78417	713/930-0350	D99
	TXD000719518	The GNI Group (Disposal Systems Inc.)	2525 Battleground Road Deer Park, TX 77536	713/930-0350	D99
	TXD027147115	Malone Service Co.	21 21st St. South Texas City, TX 77590	409/945-3301	D99 S02
	TXD055141378	Rollins Environmental Services TX Inc.	2027 Battleground Rd. Deer Park, TX 77536	713/930-2300	D80 S01/2/4 T01-4
	TXD069452340	Texas Ecologists Inc.	3 1/2 Mile S. on Petronilla Rd. Robstown, TX 78380	512/387-3518	D80 S01/2/4 T01
REGION VII	None Listed [c]				

Continued

Table 18. (Continued)

		Facility Data			
U.S. EPA Region	EPA ID Number	Facility Name	Facility Address	Telephone No.	TSD Type
REGION VIII					
Colorado	COD991300484	Rollins Environmental Services Highway 36	108555 Hwy 36 Deer Trail, CO 80105	303/386-2293	D80 S02/4 T0/2
Utah	UTD982598898	Envirocare of Utah Inc.	1 Mile S. of Clive RR Clive Site Tooele, Utah 84101	801/532-1330	D80 T S
	UTD991301748	USPCI Grassy Mountain Facility (Laidlaw Env. Services)	Sec. 16 T1 N. T12 W. Tooele, UT 84107	801/884-6841	D80 S01/2 T03
REGION IX					
California	CAD000646117	Chemical Waste Management Kettleman	35251 Old Skyline Rd. Kettleman City, CA 93239	209/386-9711	D80 S01/2 T01/2/4
	CAD990794133	Forward Disposal Site	9999 S. Austin Rd. Manteca, CA 95336	209/466-5192	D80
	CAD000633164	Laidlaw Environmental Services Imperial Valley	5295 S. Garvey W. Side Main Canl. Westmorland, CA 92281	213/830-1781	D80/83 S01/2/4 T01
	CAD980675276	Laidlaw Environmental Services (Lokern) Inc.	2500 Lokern Road Buttonwillow, CA 93206	805/762-7372	D80/83 T01 S

Continued

107

Table 18. (Continued)

			Facility Data		
U.S. EPA Region	EPA ID Number	Facility Name	Facility Address	Telephone No.	TSD Type
Nevada	NVT330010000	US Ecology Inc.	Hwy. 95 18 M.N. of Beatty, NV 89003	702/553-2203	D80 S T
REGION X					
Idaho	IDD073114654	Envirosafe Services of Idaho — Site B	Missle Base Road Grand View, ID 83624	208/384-1500	D80 S T
Oregon	ORD089452353	Chemical Waste Management of the Northwest Inc.	17629 Cedar Springs Ln. Arlington, OR 97812	503/454-2643	D80/83 S01/2/4 T01/2/4

SOURCE: (1) U.S. Environmental Protection Agency, *Treatment, Storage and Disposal Facilities with Commercial Processes*, RCRIS National Oversight Database, Washington, D.C. 20460, August 1994, and (2) U.S. Environmental Protection Agency, *The Waste System*, Office of Solid Waste and Emergency Response, Washington, D.C. 20460, 1988 — updated by 1995 telephone survey. Facilities listed in the table were identified in the U.S. EPA RCRIS National Database as RCRA disposal facilities with commercial processes; i.e., "not all processes at the site are necessarily commercial (U.S. EPA, August 1994)". For current permit and operating status, generators should contact the state/federal agency within which the disposal facility is located.

TSD Process Codes: D = disposal (nonspecific), S = storage (nonspecific), T = treatment (nonspecific), D80 = landfill, D81 = land application, D83 = surface impoundment disposal, D99 = deep well injection, S01 = container storage, S02 = tank storage, S03 = waste pile storage, S04 = surface impoundment storage, T01 = tank treatment, T02 = surface impoundment treatment, T03 = incineration, T04 = other treatment, (C) = no longer receiving hazardous waste for landfill, engaged in wastewater treatment only, and C = facility is closed or in the process of closing.

There are no commercial RCRA hazardous waste land disposal facilities listed in the U.S. EPA RCRIS (1994) database for this region.

UNIFORM HAZARDOUS WASTE MANIFEST

3. Generator's Name and Mailing Address	*SQ Generator, Inc.* *Somewhere, New England 01234*	
4. Generator's Phone *(800) 123-4567*		
5. Transporter 1 Company Name *We-Take-It Transportation Co.*	6. US EPA ID Number *NED123456789*	
7. Transporter 2 Company Name	8. US EPA ID Number	
9. Designated Facility Name and Site Address *Store Safe, Inc.* *Any Town, New England 23456*	10. US EPA ID Number *NED012345678*	

Figure 22. Generator shipment to a licensed hazardous waste storage facility.

UNIFORM HAZARDOUS WASTE MANIFEST

3. Generator's Name and Mailing Address	*Store Safe, Inc.* *Any Town, New England 23456*	
4. Generator's Phone *(800) 123-5678*		
5. Transporter 1 Company Name *We-Take-It Transportation Co.*	6. US EPA ID Number *NED123456789*	
7. Transporter 2 Company Name	8. US EPA ID Number	
9. Designated Facility Name and Site Address *Up-In-Smoke Incinerator Co.* *Sunset, New England 23456*	10. US EPA ID Number *NED012345678*	

Figure 23. Storage facility shipment to ultimate treatment facility.

transfer facilities. For information on how to locate and select treatment, storage or disposal facilities, refer to "Choosing a TSDF" in this section.

Indirect Shipments

In the case where a generator ships to an intermediate storage facility *(indirect shipment)*, the management method or identity of the *ultimate* treatment or disposal facility may not be known to the generator until a *final disposition report* is issued. Companies often initiate shipments to transfer or intermediate storage facilities when it's not economical to direct ship.

The Uniform Hazardous Waste Manifest

When a generator makes an *indirect shipment*, the generator's original manifest ends at the intermediate storage facility — the name of which is recorded in the "Designated Facility" block (9) of the Uniform Hazardous Waste Manifest; refer to Figure 22 for example. Once the storage facility accumulates a sufficient quantity of waste, a second shipment containing the generator's original material is prepared; these are often mixed shipments containing wastes from other generators headed for the same destination. At this point, however, new shipping papers are initiated and the storage facility now records its company name in the *Generator* block (3) of the manifest; see Figure 23. Although the storage facility completes the manifest as would any other generator, it is the original generator who retains "ownership" of the waste and overall responsibility for its proper management.

Final Disposition Reports

In order to ensure that wastes are properly managed, generators often designate treatment and disposal facilities that are acceptable to them and request (from the storage facility operator) periodic written reports that detail the ultimate disposition of each waste. Most reputable storage facilities and licensed waste handlers maintain computerized waste material tracking systems and will issue certified final disposition reports to generators upon request.

In the early-to-mid 1980s, generators started to demand full reports on the ultimate disposition of their wastes. In response, TSDFs began to develop paper and then computerized waste tracking systems for each shipment received. Presented below is a description (taken from sales literature distributed in 1987) of one permitted storage facility's earliest effort to offer computerized waste tracking to their customers:

> "In an era when liability is a primary concern of hazardous waste generators, waste disposal companies must be totally accountable to their customers. HAZ TRAC™ provides customers with documentation of the date, location, and technique for the disposal of every drum of hazardous waste. HAZ TRAC also offers you the ability to apply your own waste tracking codes onto the drums you ship, and have your HAZ TRAC report reflect those codes for your recordkeeping purposes. HAZ TRAC helps you to document disposal by manifest number for each pickup of waste from your facility. And at year's end, HAZ TRAC provides you with a summary of your yearly hazardous waste activity to assist you with preparation of your Generator's Annual [Biennial] Report. HAZ TRAC gives you the feedback you've been looking for!" [1]

Today, computerized TSDF waste tracking systems are the norm. Figure 24 provides one example of a fully-permitted TSD facility's "Waste Information Tracking System". With this system, Clean Harbors of Braintree, Inc. tracks each container of waste from the time it is received at their storage facility to the date that it is ultimately treated or disposed of. The

```
TSDF: CLEAN HARBORS OF BRAINTREE INC          Waste Information Tracking System                    Page    1
MAD053452637                                          CUSTOMER ABC, INC
                                                         25 MAIN ST
                                                    ANYWHERE, USA 12345
                                                       USD123456789
                                    For Manifest Dates 01/01/94 To 08/31/94 as of 09/19/94
===========================================================================================================

Line Item: 11A   Date Received: 03/09/94   Profile #: T07255   Manifest #: MAH390173   Manifested Quantity: 9   DF   495 G

Drum#       Code   Date processed   TSDF       Manifest#    Disposal method          Container Size   CC#   syscode   Empty container
206809 A32         03/26/94         CHI CLEV   MAG137118    WASTE WATER TREATMENT     55 G                   M141      CRH-ENVIROSAFE
206810 A32         03/26/94         CHI CLEV   MAG137118    WASTE WATER TREATMENT     55 G                   M141      CRH-ENVIROSAFE
206811 A32         03/26/94         CHI CLEV   MAG137118    WASTE WATER TREATMENT     55 G                   M141      CRH-ENVIROSAFE
206812 A32         03/26/94         CHI CLEV   MAG137118    WASTE WATER TREATMENT     55 G                   M141      CRH-ENVIROSAFE
206813 A32         03/26/94         CHI CLEV   MAG137118    WASTE WATER TREATMENT     55 G                   M141      CRH-ENVIROSAFE
206814 A32         03/26/94         CHI CLEV   MAG137118    WASTE WATER TREATMENT     55 G                   M141      CRH-ENVIROSAFE
206815 A32         03/26/94         CHI CLEV   MAG137118    WASTE WATER TREATMENT     55 G                   M141      CRH-ENVIROSAFE
206816 A32         03/26/94         CHI CLEV   MAG137118    WASTE WATER TREATMENT     55 G                   M141      CRH-ENVIROSAFE
206817 A32         03/26/94         CHI CLEV   MAG137118    WASTE WATER TREATMENT     55 G                   M141      CRH-ENVIROSAFE

Line Item: 11A   Date Received: 05/03/94   Profile #: T07255   Manifest #: MAH365530   Manifested Quantity: 4   DF   220 G

224043 A32         05/19/04         CHI CLEV   MAG137248    WASTE WATER TREATMENT     55 G                   M141      CRH-ENVIROSAFE
224044 A32         05/19/04         CHI CLEV   MAG137248    WASTE WATER TREATMENT     55 G                   M141      CRH-ENVIROSAFE
224045 A32         05/19/04         CHI CLEV   MAG137248    WASTE WATER TREATMENT     55 G                   M141      CRH-ENVIROSAFE
224046 A32         05/19/04         CHI CLEV   MAG137248    WASTE WATER TREATMENT     55 G                   M141      CRH-ENVIROSAFE

Line Item: 11A   Date Received: 05/19/94   Profile #: R15931   Manifest #: MAH711353   Manifested Quantity: 3   DM   165 G

229491 A31         05/23/94         SYSTECH OH MAG137325    SUPPLEMENTAL FUELS        55 G                   M141      RCL-BAY COLONY
229492 A31         05/23/94         SYSTECH OH MAG137325    SUPPLEMENTAL FUELS        55 G                   M141      RCL-BAY COLONY
229493 A31         05/23/94         SYSTECH OH MAG137325    SUPPLEMENTAL FUELS        55 G                   M141      RCL-BAY COLONY
```

Figure 24. *Waste Information Tracking System* computer printout for Customer ABC, Inc. (Courtesy Clean Harbors of Braintree, Inc.).

```
            CLEAN HARBORS OF BRAINTREE INC
                  385 QUINCY AVE.
                BRAINTREE, MA 02184

            EPA/State ID No: MAD053452637


Certificate of Treatment/Disposal - Storage and Transfer
---------------------------------------------------------

MAH390173              MAH365530              MAH711353
MAH711414              MAH711308
```


        The above described waste, received by
CLEAN HARBORS OF BRAINTREE INC (Clean Harbors)
pursuant to the manifest(s) identified above, has been treated
and/or disposed of by Clean Harbors, or another licensed facility
approved by Clean Harbors, in accordance with applicable federal
and state laws and regulations.

        Any waste received by Clean Harbors and subsequently
shipped to another licensed facility for treatment and/or
disposal has been or shall be identified as being generated
by CLEAN HARBORS OF BRAINTREE INC, in accordance with
40 CFR 264.71(c).

CLEAN HARBORS OF BRAINTREE INC

_____
By:

_____
Its:

_____
Date:

Figure 25. Certificate of Treatment/Disposal — Storage and Transfer (Courtesy Clean Harbors of Braintree, Inc.).

data set tracked by Clean Harbors, and reported back to the generator, with this program includes: TSDF waste tracking code, drum and profile sheet number (discussed in Ch. 2), receipt and process dates, incoming and outgoing manifest document number, treatment/disposal method, container type/size and empty container management information. If you haven't already done so, you should ask your TSDF for materials that describe their system. Also, request periodic reports and incorporate this information into your waste tracking program. Your ultimate goal is to document where each waste was sent to (final disposition) and how it was managed.

In summary, whenever your company ships to an intermediate storage facility, it is important that you routinely request and obtain a final disposition report from the vendor — Figures 24 and 25. In the case of direct shipments, certificates of destruction or disposal should be obtained — examples of which are provided in Figures 26 (Rollins CHEMPACK Inc.) and 27 (Laidlaw Environmental Services of SC, Inc.). These documents provide useful information concerning the ultimate disposition of your waste and will assist you in your efforts to track each unit of waste to the "end point" in the waste management cycle.

Choosing a TSDF

There are several avenues that you can take to locate and screen potential TSDFs for use by your company. Publications are available, for example, that can help you identify commercial facilities, transporters and waste handlers at the state, regional and national levels. These reference sources include:

- The *Environmental Services Directory* published by Environmental Information, LTD., 4801 W. 81st Street, Suite 119, Minneapolis, MN 55437.

- Environment Today's annual *Environmental Management Source Book* — published by Enterprise Communications Inc., 1165 Northchase Parkway N.E., Suite 350, Marietta, Ga 30067.

- O'Keefe's Regional Guides to *Environmental Services, Equipment and Supplies* published by Philip O'Keefe Co., Inc., 171 Main Street, Matawan, NJ 07747.

- *Pollution Engineering Yellow Pages,* available from Pudvan Publishing Co., 1935 Shermer Rd., Northbrook, IL 60062.

113

ROLLINS **CHEMPAK** INC.

ONE ROLLINS PLAZA
P.O. BOX 2349
WILMINGTON, DE 19899
302-426-3476
FAX 302-426-3086

October 5, 1995

ABC Company
Attention: Ms. Joan Doe
1000 North South Avenue
Pasadena, TX 77501

CERTIFICATE OF DISPOSAL

The following materials have been received and analyzed according to the Rollins Environmental Services Waste Analysis Plan (WAP) and discharged for disposal. Disposal of the material was in keeping with all applicable federal, state, and local government regulations and guidelines.

EPA ID Number
TXR000004986

Manifest Number	CHEMPAK Shipment Date	FACILITY Destruction Date
0088066	07/18/95	08/30/95

If you have any questions concerning this matter, please contact me at (302) 426-3525.

Sincerely,

Joanne R. Micallef
Rollins CHEMPAK Inc.

Figure 26. Certificate of Disposal/Destruction (Courtesy of Rollins Environmental Services).

LAIDLAW
ENVIRONMENTAL
SERVICES

HAZARDOUS MATERIALS WASTE DISPOSAL

CERTIFICATE OF COMPLIANCE AND DISPOSAL

I certify that on ___August 18, 1994___ waste material received

from ___Laidlaw Reidsville___ for work order # _100002_ , item

PW# _01286-1101_ , described on South Carolina manifest #_00005_ ,

line 11-A, was disposed of in compliance with all state

and federal laws and regulations, including 40CFR parts 260 through

268.

Facility Name: ___Laidlaw Environmental Services of SC, Inc___

Facility Address: ___Route 1, Box 255, Pinewood, SC 29125___

Facility EPA ID Number: _SCD070375985_

By: ___Sandy Wheeler___

Signature: ___Sandy Wheeler___

Title: ___Regulatory Coordinator___

Date: ___August 19, 1994___

Pinewood, SC
_____
Operations

Figure 27. Certificate of Compliance and Disposal (Courtesy of Laidlaw Environmental Services).

In general, environmental service publications/directories vary in the scope of facilities covered, data set completeness and the level of detailed information provided. While most have basic information (e.g., facility name, address and telephone number) on selected facilities, the *Environmental Services Directory*, published by Environmental Information, Ltd., contains detailed TSDF profiles with data on customer restrictions, waste types accepted, on site management technologies, regulatory status, approval time for new wastes and more. A listing of commercial facilities that reuse hazardous waste as fuel and other treatment operations (that were expected to be operational by 1993) can be found in the *Commercial Treatment/Recovery Capacity Data Set*, available from the National Technical Information Service, 5285 Port Royal Road, Springfield, VA 22161. State/regional lists of licensed facilities may also be available from state environmental protection agencies.

Screening and Evaluating TSDFs

Once facilities have been identified for potential use, they should be screened and evaluated for final selection. You should begin your screening effort by (1) contacting the company directly to obtain as much background information as possible, and (2) checking with local, state, and federal agencies that regulate the facility in question. For detailed regulatory information concerning the TSDF, you may have to either visit the agency to review the facility's inspection file or make a formal written request — under the authority of state/federal Freedom of Information Act (FOIA) laws, for example. FOIA requests to EPA headquarters can be directed to ATTN: FOI Coordinator, Office of Solid Waste and Emergency Response, U.S. Environmental Protection Agency, Washington, D.C. 20460; regional EPA enforcement divisions, however, typically have more recent and detailed information on individual facilities.

The U.S. Environmental Protection Agency offers the following advice concerning *Choosing a Hazardous Waste Hauler and Designated Waste Management Facility:* [2]

"Carefully choosing a hauler and designating a waste management facility is important. The hauler will be handling your wastes beyond your control while **you are still responsible** for proper management. Similarly, the waste management facility will be the final destination of your hazardous waste for treatment, storage, or disposal. Before choosing a hauler or designating a facility, check with the following sources:

• Your friends and colleagues in business who may have used a specific hazardous waste hauler or designated facility in the past.

- Your trade association(s) which may keep a file on companies that handle hazardous waste.

- Your Better Business Bureau or Chamber of Commerce to find out if any complaints have been registered against a hauler or facility.

- Your state hazardous waste management agency or EPA regional office, which will be able to tell you whether or not a company has a U.S. EPA Identification Number, and may know whether or not the company has had any problems.

After checking these sources, contact the hauler and designated hazardous waste management facility directly to verify that they have U.S. EPA Identification Numbers, and that they can and will handle your waste. Also make sure that they have the necessary permits and insurance, and that the hauler's vehicles are in good condition. Checking sources and choosing a hauler and designated facility may take some time — try to begin checking well ahead of the time you will need to ship your waste. Careful selection is very important."

If at all possible, and as a final step in the evaluation process, generators should visit TSD facilities to which wastes will be shipped. [Note: If your primary TSD is a commercial storage facility, you can request (for your review) a written summary of your storage facility auditor's findings concerning the ultimate treatment/disposal companies to be used.] Before visiting a TSDF, however, you should become familiar with the facility's management services and operations by reviewing company supplied literature. You should also develop an audit checklist — this will help you to remember those questions that are important to you and will allow you to gather complete and comparable data among facilities.

Examples of the types of information that could be collected during the site visit include:

1) Compliance history at the local, state, and federal levels
2) Health and safety program status and reports
3) Liability insurance/generator indemnification provisions
4) Financial status/parent organization/previous owners
5) Permits, licenses and outstanding regulatory actions
6) Acceptable/unacceptable wastes
7) History of spills, accidents and corrective measures
8) Past, current or potential future CERCLA site involvement

9) Treatment, storage and disposal methods employed and the effectiveness or integrity of those methods
10) Waste analysis program and QA/QC procedures
11) Number of employees, years of service and educational/professional training and qualifications
12) Community relations/neighbor complaints
13) Years in operation/changes in facility operations
14) Customer base serviced (e.g., federal government installations, Fortune 500 companies, etc.)

A number of reference materials exist that can help you formulate regulatory compliance questions and structure your TSD facility data collection efforts. Government handbooks such as the EPA *RCRA Inspection Manual*, the EPA *Multi-Media Investigation Manual*, and the OSHA *Field Operations Inspection Manual* are available from the National Technical Information Service (address previously provided) or from private organizations such as Government Institutes, Inc., 4 Research Place, Suite 200, Rockville, MD 20850 or Stevens Publishing Corporation, P.O. Box 2573, Waco, TX 76702.

Once the site visit has been concluded, be sure to keep a record of your findings. Throughout the evaluation process, don't be afraid to bring to the table the skills of your company's accountants and other specialists. Though you can never fully insulate your company from environmental risks, you can minimize potential future liability by carefully selecting those companies that will transport and manage your wastes.

COMPLETE MANAGEMENT COST DATA REQUIREMENTS

Overview

Complete or "fully-loaded" waste management costs were defined in Chapter 1 as the sum total of all on site and off site expenses associated with the generation and management of solid and hazardous waste. These costs include those shown in Table 19. Some of the costs shown are fixed, e.g., on site labor, and do not vary significantly by waste type. Other costs, such as off site treatment and disposal costs, however, do vary by waste type and are directly dependent upon the composition and volume of material generated. As a general rule, overall waste management costs are affected more by the chemical nature, volume and physical state of a waste than by most other factors.

Table 19. Examples of Costs Associated with the Generation and Management of Waste.

	On site Generation & Management Costs	Off Site Management Costs	
		Transportation	Treatment, Storage & Disposal (TSD)
Administrative	• Program Management • Regulatory Compliance • Vendor Contracts/Audits/Billings	• Vehicle/Equipment Fees • Demurrage Fees • Weighing Charges • State Taxes • Trailer Spotting Charges • Rejected Load Fees	• Destruction/Disposal Costs • Laboratory Fees • Out-of-Specification Surcharge • Special Handling Fees • Waste Prequalification Charge • State TSD Taxes • Repacking Service Fees
Labor	• On Site Transportation • Container/Tank Management • Maintenance/Sampling Costs • Shipment Preparation		
Materials	• Raw Material Costs, Utilities • Packaging Materials, Labels • Safety/Emergency Equipment • Containers (New/Reconditioned)		
Other	• Waste Analysis/Brokerage Fees • Generator/Toxic Material Use Fees • Equipment/Operating/Insurance • Storage Space, Penalties & Fines	Future Costs • Liability in Transportation • Potential Waste Management Facility Liability	

Incomplete Cost Accounting

Though most companies can benefit from fully-loaded waste management cost accounting, many generators do not routinely track these costs over time. Certain on site costs related to program administration (e.g., safety training costs, recordkeeping, reporting, monitoring and permitting costs, and costs associated with the coordination and supervision of shipments), operating labor and materials are often overlooked and not fully considered. It is not unusual, for example, for a department to request project funding or report on environmental expenditures without detailing the full range of costs associated with a particular waste management operation or pollution prevention project. This can leave a department or program manager at a distinct disadvantage and will result in an incomplete picture of overall program costs.

Past Practices and Trends
Historically, generators have directed most of their attention and accounting efforts toward tracking transportation, treatment, storage and disposal costs — as shown in Table 19. Since these direct costs are often more tangible, highly visible and increase at a rate that is often greater than most on site management costs, they generally attract more attention. For example, a survey of selected firms in the commercial hazardous waste management industry indicated that from 1983 to 1987, Table 20, the average charge for most management activities (not corrected for inflation) rose by at least 69 percent, with some services increasing by as much as 1,940 percent over the period.[3] A second, and more recent, cost comparison showed that hazardous waste which cost $200 to $300 per ton to dispose of in 1984, cost as much as $1,500 to $2,000 per ton in 1994.[4] Program managers, department heads and purchasing agents can easily relate to these tangible, direct costs and have all too often readily accepted them as "the price of doing business" — without fully exploring potential pollution prevention cost saving measures.

When it comes to some of the more abstract costs, such as those related to on site storage and handling or future costs associated with off site waste management, however, the picture becomes less clear. Since these costs are less tangible, the perception has often been that they are insignificant, incidental or very difficult to assess; they, therefore, have not always been factored into the overall cost for managing wastes. Though many of these costs do not readily lend themselves to precise cost accounting, experience has shown that (1) on site costs can be significant and should therefore be considered, and (2) potential future liability costs are a serious concern for many generators and should, at least, be qualitatively assessed. The General Electric Company's *Financial Analysis of Waste Management Alternatives* manual takes the future liability assessment concept one step further by pro-

Table 20
Comparison of Hazardous Waste
Management Costs Between 1983 and 1987 [5]

Waste Management Technology	Type/ Form of Waste	Price ($/gal unless otherwise noted)	
		1983	*1987*
Landfill	• 55 gal. drum	25-60/dm	64-186/dm
	• Bulk	25-90/ton	97-166/ton
Incineration	• Clean liquids, high BTU	(0.05)-0.25	1.35-2.95
	• Liquids, low BTU	0.35-1.00	1.33-3.38
	• Highly toxic liquids	1.50-3.10	2.36-5.02
Transportation		0.08-0.17/ ton-mile	0.23/ton- mile

Note: Value in parentheses denotes credit given to generator.

viding a framework for "quantifying the economic cost of future liabilities in a corporate-approved project support format."[6]

Benefits to Full Cost Accounting

When complete management costs are tracked and when they are tracked by point of generation, generators can compare and contrast selected costs by product, process or unit operation. Furthermore, when tracked by point of generation these costs can be allocated directly back to the department of origin. This helps the program manager assign fiscal responsibility and serves as an incentive to the generating department to properly manage and reduce waste. Rice[7] describes the cost allocation concept, with reference to pollution prevention, in this way: the "allocation of disposal costs is perhaps the most effective tool that can be used to heighten individual awareness of waste reduction. One very effective way to encourage such consideration is the allocation of waste disposal and handling costs to a separate budget line item, internally billed directly to the account of the generating project or organizational unit. When such costs show up as an accountable expenditure which the generator must budget and track throughout the year, additional attention is directed to them." Rice points out that accounting systems may have to be revised to accommodate such separate billing; this is true for companies "that account for waste management expenses and pollution control costs as overhead, rather than as costs incurred by distinct production processes."[8]

Cost Accounting and Pollution Prevention

Over the last several years, several documents and numerous articles have been published on the application of total cost assessment (TCA) methods to the evaluation of pollution prevention alternatives — these include the U.S. Environmental Protection Agency's (1) Office of Pollution, Prevention and Toxic Substances' 1994 *Pollution Prevention Financial Analysis and Cost Evaluation System for Lotus 1-2-3 for DOS, Version 3.4a* (for non-governmental entities, available from Tellus Institute, 11 Arlington Street, Boston, MA 02116), and (2) Office of Research and Development's 1993 *A Primer for Financial Analysis of Pollution Prevention Projects,* Washington, D.C. 20460.

TCA differs from fully-loaded waste management cost accounting in that it refers to the practice of analyzing the "costs and benefits associated with a [specific] piece of equipment or a procedure over the entire time the equipment or procedure is to be used."[9] One of the benefits of having a fully-loaded waste management cost accounting system in place is that it provides much of the cost data needed to assist in the evaluation of existing waste management practices; thereby, offering a solid foundation on which to base a total cost assessment of pollution prevention alternatives.

Finally, the U.S. EPA's standard manual of pollution prevention practice — the *Facility Pollution Prevention Guide* — contains a chapter on the "Economic Analysis of Pollution Prevention Projects." This chapter provides an overview of the TCA methodology and is reprinted in Appendix E for your convenience. Also, a comparative review of cost accounting methods as they relate to pollution prevention projects (with special reference to GE's *Financial Analysis of Waste Management Alternatives* and EPA's *Pollution Prevention Benefits Manual*) can be found in the publication titled *Total Cost Assessment: An Overview of Concepts and Methods.*[10]

Materials for Recycle

While off site management costs typically surpass on site costs for most wastes, this is not always the case. In fact when tracked, on site costs often prove to be very significant and can, in some cases, even exceed those resulting from selected off site waste management activities. This is particularly true when precious metal-bearing wastes and other valuable spent materials are involved. A good example of this is the case study presented in Ch. 3, page 39. In that case, raw material purchase costs for virgin Freon amounted to $2,200 per drum. When spent, the material retained much of its original value, off site waste management facility fees were not assessed and a $100 per drum credit was issued to the generator. Therefore, direct out-of-pocket off site waste management costs would be essentially limited to transportation, laboratory and special handling fees; of course, potential liability in transportation and handling would still be an issue.

When justifying a new project, a company in this situation would typically focus most of its accounting efforts on raw material purchase costs and other on site costs such as utilities, equipment/operating expenses and potential penalties and fines. A complete shift in emphasis from off site to on site cost accounting now takes place as raw materials used in the process and associated production and management costs are carefully considered.

Recent Trends

In recent years, there has been a general movement toward improved methods for cost accounting, especially as it relates to pollution prevention projects. At the federal level, generators across the country are required to report on specific reductions achieved in waste generation. In addition, some states have enacted legislation that requires companies to report the total costs associated with toxics use and/or waste reduction. In Oregon, for example, state law [OAR 340-135-050 (3) (d)] specifies that a "cost accounting system must be included in each plan for reducing toxic substances and hazardous wastes. The purpose of the cost accounting system is to help justify the reduction options by identifying typical costs such as:

- Purchase of toxic substances
- Hazardous waste disposal
- Toxic substance handling/storage
- Hazardous waste handling/storage
- Hazardous waste treatment
- Environmental liability and the potential risks of accidents/spills
- Compliance and regulatory costs" [11]

Whether or not you are in a state that requires the submission of cost data, keeping detailed records on total waste generation and management costs simply makes good business sense. Recognizing the importance of tracking *both* on site and off site waste management costs, each individual environmental professional must determine the level of detail necessary to achieve success in his or her hazardous waste cost accounting program.

Data Collection

Though data collection efforts will undoubtedly vary from company to company, from a practical perspective, the question becomes mostly a matter of which costs to track routinely and how to group them. The shipment journal spreadsheet (shipment journal), presented in Chapter 6, provides a practical and flexible framework for tracking and collecting cost data. Before you begin your data collection efforts, however, you should first develop specific cost categories that you intend to track routinely over time. Select those costs that are practical for you to track and group them into well defined categories. Be sure to balance your accounting efforts properly by considering both on site, as well as off site waste management costs; see examples provided in Table 19.

In order to complete the cost accounting columns of the shipment journal (Ch. 6), you will need to estimate and acquire hard data on the total costs associated with managing each waste. From the time of generation to the time of shipment, you should be able to come up with a fairly accurate estimate of on site waste management costs. Staff-hours required to complete shipping papers, inspect storage areas, load transport vehicles, manage, mark, and label containers can all be estimated without much difficulty. Other costs such as liability and off site future costs are not as easily determined but can be estimated, when desired, using the cost accounting methods described in the publications cited earlier. Also, in certain cases, some off site waste management costs may not be completely known until you or your accounts payable department receives the vendor's invoice. In other words, your purchase order or requisition may not always match the vendor's bill. This is because the bill may reflect additional costs, such as surcharges for off-spec material or repackaging fees, which could not have been entirely anticipated. Upon payment, however, you will then be in the position to record actual cost data and then calculate fully-loaded costs associated with the management of each waste.

REFERENCES

1. **Northeast Solvents Reclamation Corporation,** *Hazardous Waste Tracking* (Fact Sheet), 221 Sutton Street, North Andover, MA 01845. 1987.
2. **U.S. Environmental Protection Agency,** *Understanding the Small Quantity Generator Hazardous Waste Rules: A Handbook for Small Business,* EPA/530-SW-86-019, Office of Solid Waste and Emergency Response, Washington, D.C. 20460. September 1986.
3. **Forcella, D.,** *Volume I: The Role of Waste Minimization,* National Governors' Association, 444 North Capitol Street, Washington, D.C. 20001-1572. 1989.
4. **Egan, P.W., Enander, R.T. and Gouchoe S.,** EPA's Guidance on Waste Minimization Outlines Program Elements, in *Journal of Environmental Regulation,* John Wiley & Sons, Inc., 605 Third Avenue, New York, N.Y. 10158-0012. August 1994.
5. **op. cit. Forcella.**
6. **MacLean, R. M.,** *Financial Analysis of Waste Management Alternatives,* General Electric Company, 3135 Easton Turnpike, Fairfield, CT 06431. 1987.
7. **Rice, S.C.,** Incorporating Waste Minimization into Research and Process Development Activities, in H.M. Freeman, ed., *Hazardous Waste Minimization,* McGraw-Hill Publishing Company Inc., 1221 Avenue of the Americas, New York, New York 10020. 1990.
8. **Federal Register,** 56, 7849, 1991.

REFERENCES (Continued)

9.   **U.S. Environmental Protection Agency,** *A Primer for Financial Analysis of Pollution Prevention Projects,* Office of Research and Development, Washington, D.C. 20460. April 1993.

10. **Becker, M. and White, A.L.,** *Total Cost Assessment: An Overview of Concepts and Methods,* Tellus Institute, 89 Broad Street, Boston, MA 02110. September 1991.

11. **Oregon Department of Environmental Quality,** *Benefitting from Toxic Substance and Hazardous Waste Reduction: A Planning Guide for Oregon Businesses,* Portland, Oregon. October 1990.

Chapter 6

TRACKING OFF SITE SHIPMENTS
AND FULLY-LOADED WASTE MANAGEMENT
COST ACCOUNTING

INTRODUCTION

Chapter 5 presented an overview of three categories of data and information — manifest/point of generation data, TSDF final disposition reports, and complete management cost data — that are needed to complete your point of generation tracking and cost accounting program. In this final chapter, guidance is given on how to organize those data into a rather simple spreadsheet accounting system. The method presented provides a basis for (1) tracking each unit of waste and multiple sets of interrelated data across the entire management cycle, and (2) allocating "fully-loaded" waste management costs directly back to the product, operation or department of origin. Guidance is also provided, later in the chapter, on the analysis and presentation of waste management data. Some final thoughts are then offered as concluding remarks.

SETTING UP A TRACKING AND ACCOUNTING
SYSTEM THAT IS RIGHT FOR YOU

The U.S. Environmental Protection Agency has estimated that 250 million metric tons of hazardous waste is generated nationally each year. Of the more than 270,000 establishments[1] that generate this waste, it is likely that fewer than one in three have in place record management systems that are capable of tracking each "unit of waste" and "fully-loaded" costs across each step in the waste management cycle — i.e., from the point of generation to *final* treatment and/or disposal. Though it's a relatively simple task for most generators to track shipments of waste in aggregate, few can easily relate specific waste quantities, final treatment and disposal methods, and complete management costs directly back to the originating process or unit operation. Those companies that have in place, or are willing to develop, point of generation tracking and cost accounting systems are best prepared to meet existing, as well as future, environmental challenges.

The materials presented in this section describe a simple spreadsheet accounting method that can be used by generators of all sizes as a guide in the development of shipment and cost data tracking systems. Paper spreadsheet tracking and accounting systems are appropriate for companies with few waste streams.   Large generators with many different types of waste may be better served by self/custom-designed electronic spreadsheet or commercial computer programs that can accomplish the objectives presented below in a more convenient fashion (see the "Electronic Systems" section of this chapter).

SHIPMENT JOURNAL SPREADSHEET

Organizational Overview

The key toward developing a practical in-house accounting system is the organization of selected waste management and cost data by generating source. The general concept here is to select the types of data that you wish to track and then arrange or record them in a functional, easy to follow format. One of the simplest ways to accomplish this goal is to construct a shipment journal, or specially designed spreadsheet, in which data are tabulated chronologically, organized by point of generation —with special reference to final disposition and complete management costs. The purpose of the shipment journal is to record key data in a way that facilitates waste tracking and fully-loaded waste management cost accounting.

Constructing the Solid and Hazardous Waste Shipment Journal

Shipment journals can be developed using (1) standard accounting ledger sheets available from most office supply stores (paper system), (2) a computer, equipped with a draw/graphics software program and laser output capabilities, for customized form design (paper system, Fig. 30-34), or (3) a commercial computer software package (electronic system).

Generally speaking, shipment journals should be designed to track only the most useful information; i.e., information necessary to understand material movement and associated costs throughout the waste management cycle. Also, spreadsheet data should be organized chronologically by shipment date and by point of generation.

Journal Example: Shipment Data Tracking
Figure 28 represents a simple spreadsheet that allows for waste tracking by point of generation across each step in the waste management cycle. Note that only the most important information has been included in the sample spreadsheet. Letter and numerical codes have been assigned to identify Transportation (T) companies and Designated (interim storage) Facilities (DF), as these are only of secondary importance. Where the designated facility (as recorded on the manifest) is not an intermediate

storage facility, a DF code is not assigned and the name of the ultimate treat-ment/disposal facility is recorded in the far right hand column (see Lines 5 and 7). In the example provided, the final treatment/disposal facility for each waste is known since the generator requested and received a written report, from the intermediate storage facility (Figure 28, Lines 1 through 4), that detailed the ultimate disposition of each waste. An important feature of the sample spreadsheet is that all wastes are recorded separately by point of generation. Though similar wastes are often added together and recorded as one-line item on a Uniform Manifest (as would be the case with wastes MN-001 and MN-002, Figure 28), the integrity of this tracking system is maintained by recording each unique waste separately by generating source.

When wastes are tracked as shown in Figure 28, integrating a cost accounting function into the spreadsheet design is rather simple. First, and as discussed earlier, each generator must develop cost categories that are ap-propriate for the routine collection of data at his/her facility. Second, these cost data must then be organized by point of generation.

Journal Example: Cost Accounting
To recap from Chapter 5, in order to track complete management costs, you will need to estimate and acquire hard data on the total costs associated with managing each waste. From the time of generation to the time of shipment, you should be able to come up with a fairly accurate estimate of on site waste management costs. Off site management costs should be readily available to you in the form of itemized bills/vendor invoices.

Figure 29 is shown as a continuation sheet of the spreadsheet that was started in Figure 28. As illustrated in Figure 29, fully-loaded waste man-agement costs are tracked by point of generation. In this example, the gen-erator selected the following cost categories or column heads: On Site Man-agement Costs (i.e., Administrative/Materials/Labor/Other), Transporta-tion/Demurrage, Off Site Treatment, Storage & Disposal Costs, State Taxes, Off-specification Surcharges, Laboratory Waste Analysis Fees (including material pre-qualification charges) and Lab Pack/Repackaging Service Costs. [NOTE: The cost categories shown in Figure 29 are provid-ed for discussion only. All on site waste generation and management costs have been lumped together in the On Site Management Cost column. In practice, however, you may wish to track administrative, material, operat-ing labor or other on site cost categories separately.]

In the next to the last column, total or fully-loaded costs for managing each unique waste are summed by "Point of Generation." Following Line #1 from Figure 28 for "Waste oil (fork truck maintenance)" to the "Cost by Point of Generation" column in Figure 29, one can determine rather quickly that it cost $1,109.25 to manage 495 gallons of waste oil; or about $2.24/ gallon when all costs are accounted. Also, total waste management costs for the first three shipments of the year, from February 2 (Line #1, Figure 28)

Figure 28. Example of a simplified solid and hazardous waste shipment journal spreadsheet.

NO.	DATE	SOURCE	WASTE DESCRIPTION	WASTE NO.	CONTAINER	QTY.	T	DF	FINAL DISPOSITION
1	2/2/95	MN-001	Waste oil (fork truck maintenance)	D008	9 DM	495 G	A	1	Oil Recovery Inc. AZD456789123
		MN-002	Waste oil (grinding machines)	D008	6 DM	330 G			Oil Recovery Inc. AZD456789123
		MN-003	Spent TCA cold cleaning ops.	F001	10 DM	550 G			Recycling Ltd. NJD012345678
		WH-012	Potassim Cyanide Obsolete Inventory	P098	7 DM(30g)	210 G			Products Reuse Inc. NHD112345678
2	3/14/95	CP-001	Waste acetone from reactor cleanout	F003	20 DM	1100 G	B		Kiln Conversions Co. SCD356789123
		PH-001	Waste MEK solvent	F005	50 DM	2750 G			Kiln Conversions Co. SCD356789123
3	4/28/95	WT-001	Waste water treatment sludge	D007	65 DM	3575 G	B		Solidification Systems NYD656789123

Key: A = We-Take-It Transport Specialists, Inc., Somewhere, New England
B = Safe Way Trucking Co., Anywhere, NE
1 = Store Safe Waste Transfer Station, Inc., Any Town, NE

NOTES: NO. = Shipment number, 1 indicates first shipment of the year; DATE = Shipment date (from manifest); T = Licensed Transporter, keyed to lower left-hand corner of spreadsheet; DF = Designated Facility, as recorded on manifest and keyed to lower left-hand corner of the spreadsheet.

Figure 29. Cost accounting continuation sheet example.

	On Site Mngt. Costs	Transp./ Demurrage	TSD Costs	Taxes	Off-spec. Charges	Analytical Fees	Lab Pack/ Repack Fees	Cost by Point of Gen.	Running Sum
1	450.00	$45.00	$585.00	$29.25	-	-	-	$1,109.25	$1,109.25
2	125.00	51.00	1,590.00	79.50	-	-	-	1,845.50	2,954.75
3	2,000.00	85.00	500.00	125.00	$260.00	-	-	2,970.00	5,924.75
4	250.00	59.50	350.00	17.50	-	$100.00	$200.00	1,137.00	7,061.75
5	1,200.00	140.00	350.00	55.00	-	-	-	1,745.00	8,806.75
6	3,000.00	350.00	875.00	137.50	145.00	110.00	-	4,617.50	13,424.25
7	4,200.00	450.00	21,125.00	1,056.25	-	-	-	26,831.25	40,255.50

NOTE: On site management costs include program administration, operating labor, material costs and special fees less waste analysis and lab pack fees.

through April 28, 1995 (Line 7, "Running Sum" column Figure 29), amounted to $40,255.50.

DETAILED RECORDKEEPING GUIDANCE

Introductory Comments

Figures 28 and 29 provide a working example of a simplified spreadsheet tracking and accounting system. Data presented were used to highlight selected organizational and recordkeeping features — some of which will be further discussed below. In Figures 30 through 34, a more detailed recordkeeping framework is presented. The work sheets shown in these figures can be used as a template, guide or starting point for your own ideas in the design of your site specific tracking and accounting system.

For generators who opt to use a manual tracking and accounting system, the work sheets presented in Figures 30 through 34 can be modified to match program needs and then used as a stand-alone recordkeeping system. In other cases, where a computer spreadsheet software program or custom designed electronic waste tracking system will be used, the work sheets may serve as a reference point for system design or may be modified and used to collect or organize data for input into the program. Though the detailed guidance presented below is in reference to the organizational framework presented in Figures 30 through 34, the overarching recordkeeping principles are generally applicable to both electronic and manual recordkeeping systems.

The Sample Journal

Detailed Guidance
The work sheets shown in Figures 30 to 34 are designed so that important waste management and cost data are chronologically recorded by point of generation. The main body of the sample spreadsheet is a matrix comprised of twenty-four rows and twenty-one columns. Any one of the more than five hundred individual data cells may be quickly and conveniently accessed by Line # and column header. In this example, eleven columns (marked A through K) are non-descript and may be tailored to meet site specific needs. Data entry blocks, located along the bottom portion of the journal, are provided for recording transporter, designated facility, column totals and other key information.

Beginning Note
As discussed in Chapter 5, in order to maintain a spreadsheet tracking and cost accounting system it will be necessary to obtain information and data from a variety of sources. In some cases, time estimates (e.g., program administration and operating labor) will have to be made in order to track on site management costs. In most cases, however, all the data that you will

131

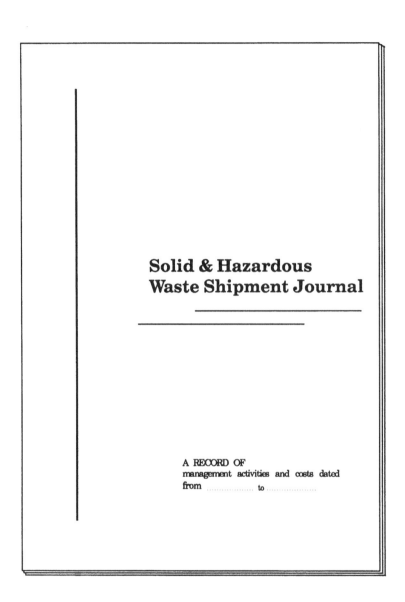

Figure 30. Example of a custom designed solid and hazardous waste shipment journal spreadsheet (paper system).

Solid & Hazardous Waste Shipment Journal

Line #	Ship. No.	DATE (MO/DY/YR) MANIFEST DOC. NO.	SOURCE/ID NO.	DESCRIPTION OF WASTE	US EPA/STATE WASTE NOs.
1					
2					
3					
4					
5					
6					
7					
8					
9					
10					
11					
12					
13					
14					
15					
16					
17					
18					
19					
20					
21					
22					
23					
24					
		TRANSPORTER		**DESIGNATED FACILITY**	

Figure 31. Start of the waste tracking section of the journal. Designed for (1) 81/2 x 11 in. 70 lb. paper stock, (2) double sided printing with Figure 30, and (3) three-hole punch along right-hand margin for storage in a three-ring binder with sections shown in Figures 32 through 34.

133

CONTAINER NUMBER/TYPE	DESIGNATED FACILITY / TRANSPORTER / AMOUNT OF WASTE	▼	▼	ULTIMATE TREATMENT/DISPOSAL FACILITY	A	Line #
						1
						2
						3
						4
						5
						6
						7
						8
						9
						10
						11
						12
						13
						14
						15
						16
						17
						18
						19
						20
						21
						22
						23
						24

YEAR .... Page .... of ....

Figure 32. Second half — waste tracking section of the journal. Designed for double sided printing with Figure 33 and three-hole punch along left-hand margin.

Solid & Hazardous Waste Shipment Journal (Continuation Sheet)

Line #	B	C	D	E	F	G	
1							
2							
3							
4							
5							
6							
7							
8							
9							
10							
11							
12							
13							
14							
15							
16							
17							
18							
19							
20							
21							
22							
23							
24							
TOTAL THIS PAGE							
TOTAL TO DATE							

FORM No. 900728/4-2

Figure 33. Reverse side — start of the cost accounting section of the journal.

135

Printed in U.S.A.

H	I	J	K	Line #
				1
				2
				3
				4
				5
				6
				7
				8
				9
				10
				11
				12
				13
				14
				15
				16
				17
				18
				19
				20
				21
				22
				23
				24

TOTAL THIS PAGE   Notes:

TOTAL TO DATE

Figure 34. End of the journal's cost accounting section. Designed for double sided printing with Figure 31 and three-hole punch along left-hand column. Copying sequence to continue without Figure 30.

need will be available in the form of shipment manifests, accounts payable records, and TSDF final disposition reports.

As a practical matter, on site and off site management and cost data are entered into the journal at the time they are first available. Entries are made chronologically at the time of each shipment. Information relating to the final disposition of each waste is recorded systematically upon receipt.

Line Numbers (#)

Because the sample shipment journal (Figures 30 through 34) is, in effect, a two-sided tabloid, line numbers are very useful points of reference. In Figures 28 and 29, for example, one can quickly scan the entire journal (Line #3) for all data relating to the management of Spent 1,1,1 Trichloroethane (MN-003). In this example, 10 drums or 550 gallons of MN-003 were shipped to Store Safe Waste Transfer Station, Inc. on February 2, 1995. At some later date (as recorded in Store Safe's monthly report to the generator, see Ch. 5 "TSDF Final Disposition Reports"), Store Safe then shipped this material to its final destination — Recycling Ltd., Blendaway, NJ. The total or fully-loaded waste management costs for these ten drums of spent solvent amounted to $2,970.00 (Line #3, "Cost by Point of Generation" column, Figure 29).

Shipment Number (Ship. No.)

The first column of the spreadsheet allows you to track the number of shipments per unit time. For the first off site shipment of each year, the number "1" would be entered in the space provided. Each successive shipment would then be numbered in ascending order. In Figure 28, 3 shipments were made between February 2 and April 28, 1995.

Shipment Date and Manifest Document Number

The second column is divided into two parts: the SHIPMENT DATE and MANIFEST DOCUMENT NUMBER (DOC. NO.). This data is obtained from the Uniform Hazardous Waste Manifest.

SHIPMENT DATE — All shipments should be recorded chronologically. Enter the month, day and year that the shipment was initiated. The record year is also entered in the upper right-hand corner of the journal (Figure 32).

MANIFEST DOCUMENT NO. — Enter either (1) the unique 5-digit, serially increasing number that you assign to each manifest, or (2) the pre-printed State Manifest Document Number assigned to each manifest by the issuing state agency.

Chronological recordkeeping allows for the rapid retrieval and analysis of waste management and cost data. Five (5) digit manifest document numbers are assigned sequentially and indicate the total number of shipments initiated over time. The State Manifest Document Number provides positive identification of the shipment and is an alternate means to access data.

Source/ID Number (No.)

The Source Identifier or Waste ID No. is the cornerstone of the entire hazardous waste tracking and cost accounting system. It is the key to accessing process, waste management and cost data.

Very often similar wastes are added together and recorded as one line item on the Uniform Hazardous Waste Manifest. The U.S. DOT proper shipping name "Waste Flammable Liquid, N.O.S.," for example, can be used to describe a mixture of wastes which actually originated from several different sources. The federal regulations allow this under certain conditions when shipping chemically compatible wastes.

In order to track each waste independently by point of generation, a separate journal entry must be made for each distinct waste stream. For example, a bulk shipment containing a mixture of 1,000 gallons of methyl ethyl ketone, 1,500 gallons of acetone, and 2,500 gallons of isopropyl alcohol, each from a different generating source, would be entered into the journal spreadsheet as three separate line items, rather than one:

Correct Entry	
PROD-02	Spent Methyl Ethyl Ketone (MEK)
POLL-06	Spent Acetone
CHEM-09	Isopropyl Alcohol, Obsolete Inventory

Incorrect Entry	
MIXTURE:	Waste Flammable Liquid, N.O.S.
PROD-02	Spent MEK, Acetone and IPA
POLL-06,	
& CHEM-09	

Description of Waste

The fourth column of the sample spreadsheet is reserved for recording the waste description assigned by you (discussed in Ch. 2).

U.S. EPA/State Waste Number(s)

For each listed and characteristic waste, enter the four-character U.S. EPA Hazardous Waste Number from 40 CFR Part 261 or appropriate state code. Enter all codes that apply to a single waste.

Container/Amount of Waste

Enter the number of containers of each waste and the appropriate abbreviation for the type of container used:

DM = Metal drums, barrels, kegs   TP = Tanks, portable
DW = Wooden drums, barrels   TT = Tank trucks
DF = Fiberboard, plastic drums   TC = Tank cars
DT = Dump truck   CY = Cylinders
CM = Metal boxes, cartons, roll-offs CW = Wooden boxes
CF = Fiber/plastic boxes, cartons   BA = Burlap, cloth bags

In the AMOUNT OF WASTE column, enter the total quantity of each waste shipped. Enter the appropriate abbreviation for the unit of measure used:

G = Gallons   L = Liquids   P = Pounds
T = Tons   Y = Cubic Yards   K = Kilograms
M = Metric Tons   N = Cubic Meters

Transporter Column

In the TRANSPORTER column, enter a code that will represent the transportation company used to ship the waste; refer to example provided in Figure 28. The block titled TRANSPORTER located in the lower left-hand corner of Figure 31 is then used to key the code to the actual name, location and U.S. EPA ID Number of the transportation company. In Figure 28, the letter "A" (Line #1/Transporter column) is used as a code for We-Take-It Transport Specialists, Inc. of Somewhere, New England.

Designated Facility and Ultimate Treatment/Disposal Facility Columns

The DESIGNATED FACILITY column is used to note intermediate storage facilities to which shipments are manifested. In Figure 28, the number "1" is entered as the facility code for Store Safe Waste Transfer Station, Inc., Any Town, New England. Upon receipt of the final disposition report, the facility name is recorded in the ULTIMATE TREATMENT/DISPOSAL FACILITY column of the journal. When a direct shipment is made, the DESIGNATED FACILITY column is left blank and the treatment/disposal facility name, location and U.S. EPA ID Number are recorded in the column titled ULTIMATE TREATMENT/DISPOSAL FACILITY; example provided in Figure 28, Lines 5 and 7.

By reviewing Figures 31 and 32, it is easy to see how unit tracking is effectively accomplished throughout each step in the waste management cycle. When the two forms are viewed side-by-side, the entire life cycle of each waste can be traced from its point of generation to ultimate treatment and/or disposal. Unit tracking continues uninterrupted and does not end at "designated facilities" when shipments are manifested to intermediate storage facilities or waste transfer stations.

Columns A through K: Fully-loaded Waste
Management Cost Accounting

In Figure 32, column A can be used to record (1) a running sum of the total amount of waste generated and shipped off site over time, or (2) waste management cost data.

All remaining non-descript columns B through J have been provided in Figures 33 and 34 for cost accounting. As illustrated in Figure 29, once data are properly organized it is a relatively simple task to track fully-loaded waste management costs by point of generation.

Finally, column K, the largest of the non-descript columns, may be used to record information unrelated to cost. For example, column K could be used to track waste minimization/pollution prevention progress or specific data on methods of destruction or disposal.

Tracking Additional Information

In designing your site-specific waste tracking and cost accounting system, you may decide that additional detail is needed. For example, you may wish to modify your waste identification numbering system (discussed in Ch. 2) to include individual drum numbers. Drum or container numbers can be used, for example, to track accumulation start dates or the contents of labpacks. [Note: A labpack is a container, e.g., 55-gallon metal drum, in which smaller containers of chemically compatible materials are over-packed. Individual drum numbers are typically assigned to each labpack by a representative of the organization performing the service. This number is then recorded on the manifest and is used to cross-reference a detailed packing list which specifies the contents of each drum.] With regard to tracking accumulation start dates, an additional spreadsheet column would be needed to record the dates that containers were filled — indicating the start of the 90 day storage clock. Finally, columns may be added to your spreadsheet to track other information or data that you feel is important.

ELECTRONIC SYSTEMS

Computerized waste tracking and cost accounting systems are more appropriate for large generators with numerous waste streams; although simple spreadsheet software programs are widely available and can be used by generators of all sizes. Dow Chemical Company's Pittsburg, California facility is one example of a large generator that has benefited from a fully automated waste tracking system. At the Pittsburg facility, Dow employs a customized computer waste tracking system that runs on a VAX computer network, equipped with bar code scanners and label printers for waste containers (see Figure 35 for a description of this system). Advantages of the system employed by Dow are many and include increased program

Through a combination of daily activities and use of computers, plant personnel monitor the status of each drum of hazardous waste generated, from the time the first drop of waste enters the drum until the sealed drum leaves the plant as part of a shipment to a RCRA-permitted TSDF. The drum tracking system allows over 99% of all drummed hazardous waste to be managed and shipped for disposal in less than 90 days. The journey of a waste drum as it proceeds through this system is described below.

Hazardous waste regulations require that a container in hazardous waste service be labeled with the name and description of the waste and the accumulation start date, in addition to other parameters. When a plant operator needs to fill a drum with waste, he or she first enters the prescribed information into the waste drum tracking program on the facility's mainframe computer. The program then generates (at a printer near the computer terminal) labels that satisfy regulatory requirements for hazardous waste and U.S. Department of Transportation requirements. Also, the label includes a bar code, which will be explained later, for use in tracking the drum.

When the label is generated the "90-day storage clock" starts ticking. The drum is filled with the waste material, solidified, sealed, and moved to a staging area, where it is inventoried by a plant operator using a hand-held laser bar code reader/data logger. These inventory data are downloaded from the data logger to the mainframe computer, and the computer database is updated to reflect the drum's status and location. Summary reports that detail drum status are readily available from the system and show the number and location of waste drums throughout the plant site, the type of waste and the generating unit, as well as the drum age. If the age of any drum exceeds 60 days the system automatically sends an alarm via electronic mail so that the drum can be located.

Twice a week, personnel from the environmental operations (EO) department make rounds of the plant and pick up waste drums for consolidation, palletizing, and banding in a centralized waste drum warehouse. Before EO personnel will accept a drum for pickup, they inspect each drum for proper solidification, labeling, cleanliness, and weight. If these criteria are not met, the drum is flagged for attention, rejected, and is not picked up, and data regarding the rejection are entered into the data logger and downloaded to the mainframe computer. Rejected drums are flagged by the computer system for special attention. If all criteria are met, the drum is again inventoried using the bar code reader and the data down-loaded to the mainframe computer. At that time, the computer database is notified that "ownership" of the inspected and accepted drum passed to EO, which is now responsible for final packaging and shipment to the TSDF.

While the drum handling process is taking place out in the plants, a facility waste scheduler makes arrangements with the TSDF for drum shipments well in advance of the actual shipping date using the information contained in the drum tracking computer database. The scheduler also provides the TSDF with a short-range forecast of waste inventory for future scheduling purposes.

When the requisite number of drums required for shipment is accumulated in the centralized waste drum warehouse, a manifest is filled out for the shipment. The drums are loaded into properly placarded vehicles operated by a licensed hazardous waste hauler, and a final inventory is taken using the bar code reader. At this time, a sticker

Figure 35. Dow Chemical Company's computerized hazardous waste tracking system. Reproduced by permission of the American Institute of Chemical Engineers. Copyright © 1992 AICHE. All Rights Reserved.

showing the manifest number for the shipment is placed on the main label of each drum. The inventory data are then downloaded one last time into the mainframe computer and the drum identification numbers are electronically transferred to a "dead drum" file, which indicates that a shipment has been made and that the drums are no longer on site.

When the drums have been shipped off site, a summary report is sent via electronic mail to each superintendent whose plant generated the wastes in the shipment. This report notifies them that a shipment was made, tallies the number of drums of waste shipped, and presents them with a cost breakdown for the shipment. This report documents regulatory compliance. Furthermore, it provides added incentive to minimize waste because each plant superintendent knows exactly what his or her waste generating activity costs.

Figure 35. (Continued)

efficiency, 90-day drum storage tracking for regulatory compliance, and an electronic mail cost allocation and reporting system. Though the cost for this system was in the "tens-of-thousands of dollars", Heilshorn and Mac-Dougall, of the Dow Chemical Co.,[2] state that other generic, commercially available computer tracking programs are available for far less.

For generators who wish to computerize their waste tracking system, and have relatively simple needs, computer spreadsheet programs (e.g., Microsoft Office, Lotus 1, 2, 3) can be purchased for a few hundred dollars. These software programs can be used to design a spreadsheet tracking and cost accounting system as described herein; generators who make frequent shipments of multiple wastes may find that computer spreadsheet programs offer a powerful advantage when it comes to data management/analysis and report generation. Software programs for personal computers can also be used to create electronic versions of the forms presented in this book, as well as store and print out formated data for daily use.

Commercial programs are available for purchase from a variety of sources. Software written specifically for chemical and hazardous waste tracking/management are often advertised or listed in trade journals and related environmental publications. Environmental Protection's annual *Hazardous Waste Software Guide* (published by Stevens Publishing Corp., 3630 IH-35, Waco, TX 76706), Chemical Engineering Progress' annual *Software Directory* (published by the American Institute of Chemical Engineers, 345 E. 47 St., New York, NY 10017), Environment Today's *Environmental Management Source Book* (publisher's address provided in Ch. 5) and Pollution Equipment News' annual *Buyer's Guide* (publisher's address previously provided) are four examples.

In the final analysis, the decision of whether to use a manual, partially automated or fully computerized waste tracking and cost accounting system is yours. You should not only feel comfortable with the system, since you will be maintaining and using it on a routine basis, but you should also make sure that it contains all of the features necessary to make your program a success.

ANALYSIS AND PRESENTATION OF DATA

Periodically you will need to sort, analyze and report on waste management and cost data that have been collected over time. Waste management and cost data are typically analyzed to assess the operational or financial status of a program. Studies, for example, may be conducted to determine trends or evaluate changes in the quantities of waste generated, total management costs, or methods used to treat or dispose of wastes. The product of your data analysis efforts will often be a written report compiled for internal use or submitted to an outside agency in partial fulfillment of your overall regulatory obligations. Whenever you report data, however, keep in mind that the conclusions drawn are only as good as the quality (or completeness) of data used in the analysis. This is why it is so important to pay close attention to detail throughout all of your data collection and recordkeeping efforts.

CHARTS AND GRAPHS

One of the most effective ways to evaluate trends and communicate findings is to construct charts and graphs. Charts and graphs are pictorial representations of data that highlight important trends and relationships. As recommended by PDS, Inc., a Florida based consulting firm, "when using charts, graphs and tables, remember to use labels and titles. Keep [them] simple, and report all the facts needed to be fair and accurate. Charts and graphs are very useful in communicating your data to others when gaining approval or support."[3]

The Pareto Chart

One type of bar chart that is very useful in the analysis of waste management data is the Pareto chart. This chart is named after the distinguished economist Vilfredo Pareto, who studied the maldistribution of wealth in Italy. His original work was later generalized and universally applied to data of all types by J.M. Juran, who coined the popular term "vital few and trivial many."[4,5] Pareto's law simply states that "results and causes of results are not equally distributed. Some call it the 80-20 rule, [where] 80% of the results are caused by 20% of the causes. By applying this concept ... small improvements on the right problems can yield significant

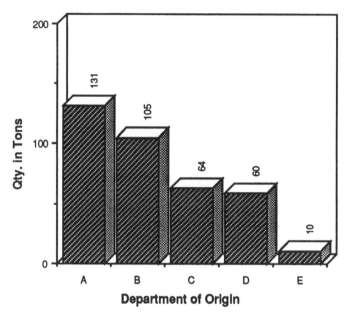

Figure 36. Pareto chart: quantity of solid waste generated in one year by department of origin[6]

increases in effectiveness."[7] The Pareto chart is one technique that is used to help separate the "vital few from the trivial many" and helps analysts focus their attention on those areas of greatest concern. When applying the Pareto principle to waste generation and management data there are two important things to remember: (1) although the pareto chart often identifies important trends in the data, the trends may not be obvious or may not be as dramatic as the 80-20 rule would suggest — therefore, further sorting and analysis may be required, and (2) the data in the chart should always be evaluated in conjunction with other pertinent information; for example, in the discussion of Figure 36 that follows, a project team considered waste management costs, chemical composition and potential liability in conjunction with data presented in the Pareto chart.

Hoechst Celanese Corporation

In order to construct a Pareto chart of waste generation or management data, all one needs to do is rank order the data set in the form of a bar or column chart. One example of the Pareto principle applied to waste generation data is given in Figure 36. In an effort to develop a structured approach to solid waste management, a team of employees at Hoechst Celanese Cor-

poration used a Pareto chart to organize waste generation data by department of origin. The Pareto chart provided the team with a visual benchmark
and direction for future waste reduction efforts. Not only did the chart show
the relative standing of each waste generating department, but it also proved
to be an effective tool for communicating project direction to upper
management. Considering management costs, potential liability and total
volume, the team decided to concentrate its resources and waste reduction
efforts on Department A. At the conclusion of the project, team members
made recommendations which led to new material handling techniques and
process changes that eventually resulted in a 78% decrease in waste output
— from 131 tons per year down to approximately 28 tons in less than ten
months.

**Bar Charts, Pie Charts, Line Graphs
and Statistical Tables**

The Pareto chart is but one variation of a bar/column chart (see Figure
37), and is one of several statistical tools that can be used to organize and
evaluate data. Four basic statistical tools that are often used to analyze and
present waste generation and management data are presented in Figure 38.
In this table, each statistical tool is listed along with the type(s) of data comparisons that are possible. Below, each of the data "comparison" techniques
listed in Figure 38 is characterized by Zelazny.[8] Several examples of each
comparison technique, as applied to waste generation and management
data, have also been prepared and follow each of Zelazny's four descriptions.

Time-Series Comparison (Column/Bar Chart, Line Graph): With
the time series comparison, the analyst is "not interested in the size of
each part in a total or how they're ranked, but in how they change over
time, whether the trend over weeks, months, quarters, years is increasing, decreasing, or remaining constant."
A column/bar chart or line graph is often used to present time series
data. For example, in the analysis of waste management data, time-
series comparisons can be used to show:

■ The total quantity of waste generated per
month, quarter or year (sample illustration
provided in Figure 38).

■ The increase or decrease (resulting from
source reduction activities, for example) in
waste management costs over time; refer to
"Line Graph", Figure 38.

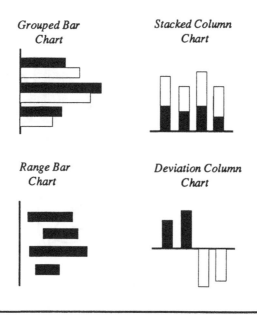

Figure 37. Variations of the bar and column charts.

■ The rise in the total number of staff-hours required to administer environmental programs

Item Comparison (Column Bar/Chart): The item comparison is used "to compare how things rank: are they about the same, or is one more or less than the others?" For example:

■ Of the five departments, Department A ranks first in total solid waste generation (see Pareto chart Figure 36).

■ The rate of waste generation from process X is much less than that of either process Y or Z.

■ SQ Generator, Inc. has shipped more waste to the Up-in-Smoke Waste Treatment facility than to any other commercial TSDF.

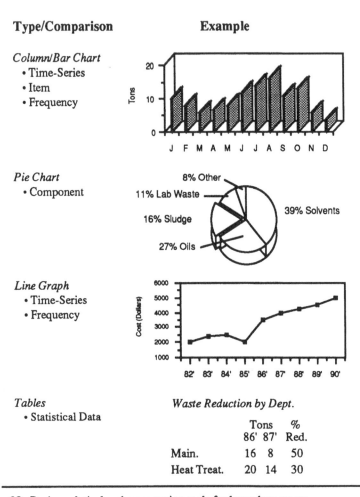

Type/Comparison	Example

Column/Bar Chart
• Time-Series
• Item
• Frequency

Pie Chart
• Component

Line Graph
• Time-Series
• Frequency

Tables
• Statistical Data

Figure 38. Basic analytical and presentation tools for hazardous waste management data.

Frequency Comparison (Column/Bar Chart, Line Graph): The frequency comparison "shows how many items fall into a series of progressive numerical ranges." For example:

■ The majority of hazardous waste shipments are made every 60 to 75 days.

■ Sixty-five percent of all chemical spills occur on the second and third shift.

Component Comparison (Pie Chart): The component comparison is principally used to show "the size of each part as a percentage of the total."

The pie chart is one of the best ways to graphically express ratio shares or component comparison data. The following are examples of hazardous waste data comparisons that can be made using a pie chart:

- The pie chart can be used to express the relative amounts of wastes generated plant-wide. As an example: For the year 1995, spent solvents (pie chart example Figure 38) represented 39% of the total of all wastes generated on site.

- Waste management cost data can be broken down into component parts and shown as a percentage of the whole. For example: Costs for on site waste management, shown as a percentage of the total, could be compared to off site transportation and disposal costs for a given period of time.

Finally, data that are used to create charts and graphs are often presented in the form of statistical tables. The usefulness of a statistical table lies in the fact that it "summarizes and often readily identifies some of the more striking features of the data."[9] The statistical table presented in Figure 38, for example, shows the relative amount of hazardous waste generated in each of two years and the percent reduction achieved. Tables are also often used to organize textual data by category or by some common variable. Whichever method(s) you choose to organize and present your waste generation and management data, be sure to "keep it simple" so that messages are clearly communicated and may, therefore, be easily understood.

CLOSING REMARKS

This book would not be complete without emphasizing the importance of source reduction as it relates to your on site waste management program. Defined by the Pollution Prevention Act of 1990 as "any practice which reduces the amount of any hazardous substance, pollutant, or contaminant entering any waste stream or otherwise released into the environment (including fugitive emissions) prior to recycling, treatment, or disposal," the term source reduction (used interchangeably with "pollution prevention") has come to symbolize a new era in environmental protection.

In recent years, several excellent books[10, 11, 12, 13] and numerous articles have been written on the related subjects of waste minimization and pollution prevention (source reduction). These and other reference materials are available to help you establish an effective pollution prevention program within your company. In addition, many state agencies, including universities, have established technical assistance programs designed to help business and industry reduce their use and disposal of toxic and hazardous materials.

By reducing waste at the source, you will not only be doing your part to improve environmental quality and worker health and safety, but you can also save your company money and reduce the number of hours that it takes you to administer and monitor on site programs. Because less waste is generated, you will spend less time recording and tracking shipments and associated waste management costs. Since waste management data are now organized by source, and since pollution prevention activities start at the point of generation, you are now in a good position to evaluate your options.

The principal goal of this work has been to help you build an effective waste material tracking and cost accounting program. As fully-loaded costs are tracked by point of generation, they should be routinely allocated directly back to the department and process or operation of origin. Though this may require a fundamental change in the way your company accounts for environmental expenditures, it will — over time — prove to be a valuable tool in your efforts to manage and control environmental costs.

In closing, the data collection and recordkeeping guidance provided herein should serve as a point of departure for other tracking and accounting initiatives, such as those that track air and water release data or total cost accounting for pollution prevention projects. Whichever direction you choose to steer your program or whatever course you chart, I hope that the information and materials presented in this book make your professional life a little easier and more enjoyable.

REFERENCES

1. Mazmanian, D. and Morell, D., *Beyond Superfailure: America's Toxics Policy for the 1990s*, Westview Press, Inc., 5500 Central Avenue, Boulder, Colorado 80301-2847. 1992.

2. Heilshorn, E.D. and MacDougall, J.J., Identify and Track Hazardous Wastes Effectively, in *Chemical Engineering Progress,* American Institute of Chemical Engineers, 345 E. 47th St., New York, NY 10017. November 1992.

3. PDS, Inc., *Roadmap to Problem Solving,* Clearwater, Florida. 1981.

4. Juran, J.M., Pareto, Lorenz, Cournot, Bernoulli, Juran, and Others, in *Industrial Quality Control.* October 1950.

5. Juran, J.M., *Quality Control Handbook*, McGraw-Hill Publishing Company, 1221 Avenue of the Americas, New York, NY 10020. 1974.

6. Enander, R.T. and Nester, D.J., A Structured Approach to Solid Waste Management, in *Chemical Engineering Progress,* American Institute of Chemical Engineers, 345 E. 47 St., New York, NY 10017. April 1988.

REFERENCES (Continued)

7. **op. cit. PDS, Inc.**
8. **Zelazny, G.,** *Say it with Charts: The Executive's Guide to Successful Presentations,* Dow Jones-Irwin. 1985.
9. **Khazanie, R.,** *Elementary Statistics In a World of Applications,* Scott, Foresman and Company, Glenview, Illinois. 1979.
10. **U.S. Environmental Protection Agency,** *Facility Pollution Prevention Guide,* Risk Reduction Engineering Laboratory, Cincinnati, OH 45268. 1992.
11. **Higgins,T.E.,** *Hazardous Waste Minimization Handbook,* Lewis Publishers, Inc., 2000 Corporate Blvd., N.W., Boca Raton, FL 33431. 1989.
12. **Overcash,M.R.,** *Techniques for Industrial Pollution Prevention: A Compendium for Hazardous and Nonhazardous Waste Minimization,* Lewis Publishers, Inc., 2000 Corporate Blvd., N.W., Boca Raton, FL 33431. 1986.
13. **Freeman,H.M.,** *Industrial Pollution Prevention Handbook,* McGraw-Hill Publishing Company, Inc., 1221 Avenue of the Americas, New York, New York 10020. 1990.

Appendix A

U.S. EPA Container Management and Storage Requirements for Generators of Hazardous Waste

USE AND MANAGEMENT OF CONTAINER STANDARDS [a, b]
Condition of Containers
1. If a container holding hazardous waste is not in good condition, or if it begins to leak, then the hazardous waste must be transferred to a container that is in good condition, or managed in some other appropriate way.
Compatibility of Waste with Container
1. The owner or operator must use a container made or lined with materials which will not react with, and are otherwise compatible with, the hazardous waste to be stored, so that the ability of the container to contain the waste is not impaired.
Management of Containers
1. Containers holding hazardous waste must always be closed during storage, except when it is necessary to add or remove waste.
2. Containers holding hazardous waste must not be opened, handled, or stored in a manner which may rupture the container or cause it to leak.
Inspections
1. Areas where containers are stored must be inspected at least weekly, looking for leaks and for deterioration caused by corrosion or other factors.
Special Requirements for Ignitable or Reactive Wastes
1. Containers holding ignitable or reactive waste must be located at least 15 meters (50 feet) from the facility's property line.
2. Precautions must be taken to prevent accidental ignition or reaction of ignitable or reactive waste. This waste must be separated and protected from sources of ignition or reaction including but not limited to: open flames, smoking, cutting and welding, hot surfaces, frictional heat, sparks (static, electrical, or mechanical), spontaneous ignition (e.g., from heat producing chemical reactions), and radiant heat.
3. Smoking and open flames must be confined to specially designated locations.
4. "No Smoking" signs must be conspicuously placed wherever there is a hazard from ignitable or reactive waste.

Special Requirements for Incompatible Wastes
1. Incompatible wastes, or incompatible wastes and materials, must not be placed in the same container.
2. Hazardous waste must not be placed in an unwashed container that previously held an incompatible waste or material.
3. A storage container holding a hazardous waste that is incompatible with any waste or other materials stored nearby in other containers, piles, open tanks, or surface impoundments must be separated from the other materials or protected from them by means of a dike, berm, wall or other device.

PREPAREDNESS AND PREVENTION [a]

Maintenance and Operation
1. Storage areas must be maintained and operated to minimize the possibility of fire, explosion, or any unplanned sudden or non-sudden release of hazardous waste constituents to air, soil, or surface water which could threaten human health or the environment.

Required Equipment
1. Facilities that store hazardous waste must be equipped with:
  a. An internal communications or alarm system capable of providing immediate emergency instruction (voice or signal) to facility personnel;
  b. A device, such as a telephone (immediately available at the scene of operations) or a hand-held two-way radio, capable of summoning emergency assistance from local police departments, fire departments, or State or local emergency response teams;
  c. Portable fire extinguishers, fire control equipment (including special extinguishing equipment, such as that using foam, inert gas, or dry chemicals), spill control equipment, or automatic sprinklers, or water spray systems;
  d. Water at adequate volume and pressure to supply water hose streams, or foam producing equipment, or automatic sprinklers, or water spray systems.

Testing and Maintenance of Equipment
1. All communications or alarm systems, fire protection equipment, spill control equipment, and decontamination equipment, where required, must be tested and maintained as necessary to assure its proper operation in time of emergency.

Access to Communications or Alarm System
1. Whenever hazardous waste is being poured, mixed, spread, or otherwise handled, all personnel involved in the operation must have immediate access to an internal alarm or emergency communication device, either directly or through visual or voice contact with another employee.

2. If there is ever just one employee on the premises while the facility is operating, he must have immediate access to a device, such as a telephone (immediately available at the scene of operation) or a hand-held two-way radio, capable of summoning external emergency assistance.

Required Aisle Space

1. Aisle space must be maintained to allow the unobstructed movement of personnel, fire protection equipment, spill control equipment, and decontamination equipment to any area of facility operation in an emergency, unless aisle space is not needed for any of these purposes.

---

a SOURCE: Title 40 of the Code of Federal Regulations, Parts 262 and 265.

b Generators who possess a RCRA Part B permit to store hazardous waste for more than 90 days must comply with 40 CFR Part 264 Subpart I container use and management requirements. Specific containment design criteria for storage areas that store containers holding liquid wastes are given in §264.175 and include: (1) a base must underly the containers which is free of cracks or gaps and is sufficiently impervious to contain leaks, spills, and accumulated precipitation until the collected material is detected and removed; (2) the base must be sloped or the containment system must be otherwise designed and operated to drain and remove liquids resulting from leaks, spills, or precipitation, unless the containers are elevated or are otherwise protected from contact with accumulated liquids; (3) the containment system must have sufficient capacity to contain 10% of the volume of containers or the volume of the largest container, whichever is greater. Containers that do not contain free liquids need not be considered in this determination; and (4) run-on into the containment system must be prevented unless the collection system has sufficient excess capacity in addition to that required in [3] to contain any run-on which might enter the system.

Appendix B

Raw Material and Product Storage Recommendations

[Author's Note: These guidelines are equally applicable to the storage of containerized hazardous waste.]

CONTAINER STORAGE GUIDELINES

- Store containers in such a way as to allow for visual inspection for corrosion and leaks.

- Stack containers in a way to minimize the chance of tipping, puncturing, or breaking.

- Prevent concrete "sweating" by raising the drum off storage pads.

- Provide adequate lighting in the storage area.

- Maintain a clean, even surface in transportation areas.

- Keep aisles clear of obstruction.

- Maintain distance between incompatible materials.

- Avoid stacking containers against process equipment.

- Use proper insulation of electric circuitry and inspect regularly for corrosion and potential sparking.

- Document all spillage.

SOURCE: U.S. Environmental Protection Agency, Facility Pollution Prevention Guide, Office of Research and Development, Washington, D.C. 20460. May 1992.

Appendix C

Toxicity Profiles for Toluene and Dichloromethane

Each profile (1) presents a summary of available toxicological and environmental data, and (2) was prepared using IRIS and ATSDR (U.S. PHS) as primary reference sources; other secondary data sources were used to a lesser extent as noted. These profiles demonstrate how secondary data sources can be used by environmental, safety or health professionals to develop written summaries that lead to a general understanding of the hazards associated with chemicals and chemical wastes. Information of this type can assist, for example, in the evaluation of human or environmental health effects of process input materials or may be used as a basis from which to communicate waste/material hazards.

TOXICITY PROFILE FOR TOLUENE

The principle route of human exposure to toluene is through inhalation. Toluene does not persist in the environment — it is volatile and rapidly degrades in soils and the atmosphere (U.S. PHS, 1994). It has low water solubility, but once released to the environment it has been reported to migrate via the air, surface water and groundwater pathways.

Toluene is a known respiratory irritant and central nervous system depressant (causing headache, dizziness, fatigue, drowsiness, incoordination, collapse, coma and death) with strong human evidence of neurotoxicity resulting from inhalation exposure to high concentrations. Deaths have been reported following acute inhalation exposure to high concentrations of toluene. Long-term human exposure to toluene in air has been associated with a range of adverse effects. Degeneration of tissues lining the nasal cavity has been demonstrated in chronic inhalation exposure studies on rats. Toluene has been shown to have chronic toxic effects on the liver and kidneys of animals via ingestion. Though limited data exists on reproductive and developmental toxicity, toluene has been suspected of causing congenital defects in cases where mothers were known to have abused or had otherwise been exposed to toluene during pregnancy (IRIS, 1995).

The U.S. Environmental Protection Agency (EPA) has determined that toluene is not classifiable, EPA weight-of-evidence designation D, as to human carcinogenicity due to the lack of human and inadequate animal data (IRIS, 1995).

BACKGROUND INFORMATION

Toluene is a clear colorless liquid with a sweet, pungent odor that is noticable in the air at a concentration of 8 ppm (U.S. PHS, 1994). In 1987, it was the 24th highest-volume chemical produced in the U.S. (Sax et al., 1987) and is derived from petroleum and coal-tar light oil (Sax et al., 1992). Toluene is a flammable, organic solvent that is used in a wide variety of applications and industrial/manufacturing processes. 89% of the toluene produced in the U.S. remains in gasoline as a nonisolated mixture of benzene-toluene-xylene; 11% is isolated by distillation and used for a variety of purposes (U.S. PHS, 1994). Toluene is commonly used, for example, in the manufacture of paint, lacquers, thinners, coatings and glue. Toluene is also used as an adhesive solvent in plastic toys and model planes and in the manufacture of gums, resins, chemicals (e.g., saccharin, medicines, dyes, and perfumes) and explosives (e.g., TNT) (Sax et al., 1987).

It has been estimated that 100,000 (Sittig, 1985) to 1.6 million (Howard et al., 1990) workers are exposed to toluene in the United States; children and hobbyists (e.g., painters, paint strippers, printers, silk-screeners, and woodworkers) are also at potential risk for exposure. Toluene's euphoria-producing effect and wide-spread availability, as an ingredient in rubber and plastic cement, has led to substance abuse ("glue-sniffing") among adults and children (IRIS, 1995; Upton et al., 1993).

PHYSICAL/CHEMICAL PROPERTIES

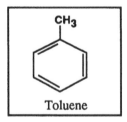

Toluene

CAS Number: 108-88-3
Molecular Formula: C7H8
Molecular Weight: 92.15
Solubility in Water: 515 mg/l, 20°C
Liquid Density: 0.866 (20°/4°C)
Vapor Density: 3.14
Vapor Pressure: 28.4 mm Hg, 25°C (Howard, 1990)
Flash Point: 40°F
Henry's Law Constant: 5.94 E-03 atm-m³/mole (Howard, 1990)

TOXICOKINETICS

Toluene has been shown to readily cross respiratory and gastrointestinal tract membranes; absorption through the skin, however, is slow. The principal route of human exposure to toluene is through inhalation. Existing data suggests that 50 - 60% of the toluene inhaled is absorbed into the body (IRIS, 1995; U.S. PHS, 1994).

Following inhalation exposure, toluene is rapidly absorbed through the lungs and into the bloodstream. Once in the blood, toluene is distributed to the brain and liver, and has been found in adipose tissue (in one study, the half-life of toluene in adipose tissue of male humans following exposure to 300 ppm ranged from 0.5 to 2.7 days) of both humans and animals, as well as in the bone marrow, spinal nerves, spinal cord and white matter of the brain in mice following inhalation exposure (U.S. PHS, 1994). Toluene is rapidly metabolized in the liver through a series of oxidation reactions leading to benzoic acid and ultimately to hippuric acid (the principal metabolite, though others include: o-cresol, p-cresol and phenol), which is then excreted in the urine (Amdur, et al., 1991). 60 to 75% of the toluene absorbed is excreted as hippuric acid within 12 hours of exposure; "much of the remaining toluene is exhaled unchanged" (U.S. PHS, 1994). Sex dependent differences in the metabolism of toluene and ethnic differences in urinary excretion of metabolites have been reported among Chinese, Turkish, and Japanese solvent workers (U.S. PHS, 1994).

No human studies involving absorption, distribution or excretion following oral exposure could be found. Results of animal studies indicate that toluene is absorbed at a lesser rate through the gastrointestinal tract as compared to the respiratory tract. Following dermal exposure, toluene was shown to absorb at a rate of 14 to 23 mg/cm^2/hr in human forearm skin (U.S. PHS, 1994). Soaking the skin of human volunteers with toluene for a short period of time (five minutes in one study) demonstrated individual differences in the rates of absorption (U.S. PHS, 1994). No studies involving the distribution or excretion of toluene following oral or dermal exposure in humans or experimental animals could be found.

MAMMALIAN TOXICITY

Most of the animal data that exists results from inhalation exposure studies. Limited animal data exists on the effects of oral and dermal exposure to toluene. Adverse effects on the central nervous system (CNS), liver and kidneys have been documented in laboratory animals.

Toluene has been shown to be a developmental toxin, but not a reproductive toxin in animal studies. Inadequate data exists to determine whether toluene is carcinogenic in rodents. Toluene was not found to be genotoxic in a majority of assays (IRIS, 1995).

Inhalation

In animals, toluene has been shown to have adverse effects on the CNS, kidneys, lungs and liver following inhalation exposure. Acute/subchronic exposure to toluene has been demonstrated to produce (1) both excitatory and depressant effects on the CNS of experimental animals, (2) changes in the levels of brain neurotransmitters in rodents, at 400 ppm/30 days of exposure for example, and (3) morphological effects on the brain at 320 ppm/30 days of exposure (U.S. PHS, 1994). Inhalation exposure has also been found to have irreversible high-frequency hearing loss and mild-to-moderate degeneration in the olfactory and respiratory epithelium of the nasal cavity in rats (IRIS, 1995).

Ingestion/Dermal

Limited animal data exists on the effects of oral and dermal exposure to toluene. Much of the oral exposure data that does exist is in the form of lethality studies. In rat studies where toluene was administered by gavage, no toxic effects on the respiratory, gastrointestinal tract or hematopoietic system were found, though significant changes in liver and kidney weights were noted (U.S. PHS, 1994).

When administered orally to mice, toluene was not found to be a developmental nor a reproductive toxin, though certain neurological functional deficits have been observed (U.S. PHS, 1994). No animal studies regarding respiratory, cardiovascular, immunological, genotoxic or carcinogenic effects could be found following oral exposure to toluene.

Animal data (rabbit and guinea pig) indicate that toluene is slightly to moderately irritating to the skin and slightly to severely (i.e., moderately severe) irritating to the eyes of rabbits (U.S. PHS, 1994). No animal studies regarding lethality, systemic, immunologic, neurologic, developmental, reproductive, or genotoxic effects could be found following dermal exposure to toluene.

HUMAN TOXICITY

Most of the human data that exists is derived from inhalation exposure studies. Deaths have been reported following acute inhalation exposure to toluene; however, quantitative data on dosage is lacking. Impaired central nervous system function is the primary toxic effect following short-term exposure. Long-term human exposure has been associated with permanent brain damage, impaired vision, speech and hearing loss, and loss of memory and balance (U.S. PHS, 1994). The U.S. EPA has determined that toluene is not classifiable as to human carcinogenicity due to the lack of human and inadequate animal data (IRIS, 1995).

Inhalation

The principal route of human exposure to toluene is through inhalation. Toluene is a known respiratory irritant and the primary target organ affected by human exposure to toluene is the brain (IRIS, 1995). Deaths have been reported following acute inhalation exposure to toluene. In England, "approximately 80 deaths per year are associated with solvent abuse (U.S. PHS, 1994)."

Human occupational/volunteer studies and case reports have shown that the toxic effects of toluene on the CNS correlate well with the amount and duration of exposure. Short-term human exposures to low concentrations can result in fatigue, confusion, general weakness and memory loss, while exposure to higher concentrations can cause drowsiness, permanent brain damage, unconsciousness and, in some cases, death (U.S. PHS, 1994). Long-term human exposure to low, moderate or high concentrations of toluene in the air has been associated with a range of adverse effects: slight effects on the kidneys resulting from long-term/low concentration exposure (though data are equivocal); and permanent brain damage, impaired speech, vision and hearing loss, loss of muscle control, loss of memory and balance, and possible developmental and immunological effects (data are equivocal) resulting from long-term/moderate to high concentration exposures (U.S. PHS,1994).

Though limited human data exist, case reports and epidemiological studies suggest that toluene may alter liver function in humans, following inhalation exposure (IRIS, 1995). In general, however, the liver and kidney are not thought to be primary target organs following human inhalation exposure to toluene (U.S. PHS, 1994). Toluene has also been suspected of causing congenital defects in cases where mothers were known to have abused or had otherwise been exposed to toluene during pregnancy (IRIS, 1995). Existing human data are inconclusive with regard to genotoxicity (U.S. PHS, 1994).

Ingestion/Dermal

Human studies regarding health effects following oral exposure to toluene are limited. Toluene may damage human skin due to its ability to dissolve skin oils (U.S. PHS, 1994). No other studies were found regarding human health effects following dermal exposure to toluene.

ECOLOGICAL TOXICITY

Toluene can readily biodegrade in fresh water, seawater, and estuarine environments. The rate of degradation is primarily dependent upon the presence of acclimated microbial populations, temperature and pH. High concentrations may be toxic to microorganisms and, at least initially, may interfere with naturally occurring biodegradation processes. Toluene will not significantly bioconcentrate in aquatic organisms, though it has been found in eels, Manila clams, mussel, algae and golden ide fish (Howard, 1990).

Toluene's high vapor pressure, low water solubility, low bioconcentration factor and rapid photo/biochemical degradation rates indicate that significant levels of toluene would not persist long enough in the environment to present a significant long-term ecological threat. Large releases of toluene to a waterway — resulting from a spill or accidental discharge for example — could, however, be acutely toxic to aquatic life and result in measurable disturbances of the receiving ecosystem. Representative marine organisms appear to be more than twice as sensitive to the acute effects of toluene as compared to freshwater species.

No data on domestic or wild animals could be found.

FATE AND TRANSPORT

Toluene has been found in drinking water (surface and groundwater supplies), industrial effluents, surface waters (both fresh and seawater), rain water, sediment/soil samples and in the atmosphere at remote, rural, urban and suburban locations.

Most of the toluene entering the environment is released directly to the air. In the atmosphere, toluene is rapidly degraded (t1/2 = 3hr to ~ 1 day) by photochemical reactions with hydroxyl radicals (Howard et al., 1990). Toluene is effectively scrubbed from the atmosphere by rain, thereby re-entering the hydrosphere, and though not subject to direct photolysis it has been shown, when complexed with molecular oxygen, to absorb light at wavelengths > 290 nm, making it potentially subject to photochemical degradation in this form (Howard et al., 1990).

Due to its low water solubility and relatively high vapor pressure, toluene releases to surface water and soil tend to evaporate rather quickly (U.S. PHS, 1994). The rate of toluene volatilization from surface water varies with turbulence (static waters t1/2 = 1-16 days; turbulent waters t1/2 = 5-6 hours), while the rate of volatilization from soil is dependent upon temperature, humidity and soil type — under typical conditions 90% of the toluene in near-surface soils can be expected to volatilize within 24 hours (U.S. PHS, 1994).

In surface water, toluene will biodegrade at varying rates, depending upon the presence of acclimated microorganisms (t1/2 = days to several weeks), but will not significantly hydrolyze or adsorb to sediment (Howard et al., 1990). Toluene has a "moderate tendency" to bioaccumulate in aquatic organisms; levels of accumulation being dependent upon several factors, including species metabolism (U.S. PHS, 1994). Toluene is not subject to direct photolysis in air or water. In soil, toluene may be subject to microbial degradation (t1/2 typical in the environment = 1-7 days; U.S. PHS, 1994) or may leach to ground water, especially in soils with low organic content, where it will not significantly hydrolyze but may be subject to slow biodegradation (Howard et al., 1990).

REGULATIONS AND STANDARDS (U.S. PHS, 1994)

Regulations
Occupational Safety and Health Administration (OSHA)
OSHA Permissible Exposure Limit: TWA 100 ppm; STEL 150 ppm

Resource Conservation and Recovery Act
U.S. EPA Listed Hazardous Waste:
F005 Spent non-halogenated solvent (ignitable, toxic)
U220 Toxic waste

CERCLA/Superfund
Reportable Quantity for Accidental Release: 1000 lbs.

Food and Drug Administration
Residual Solvent in Finished Resins

Guidelines
Worker Exposure Guidelines
ACGIH TLV: TWA 100 ppm; STEL: 150 ppm
NIOSH: TWA 100 ppm; STEL: 150 ppm

Safe Drinking Water Act
Maximum Contaminant Level Goal: 1 mg/l

Drinking Water Health Advisory
1 day (child): 2E+1 mg/l

Clean Water Act
Ambient Water Quality Criteria for Human Health
Water & Organisms: 1.43E+4 ug/l
Organisms Only: 4.24E+5 ug/l

164 Toxicity Profile

REFERENCES (Toluene)

Amdur, M.O. et al., *Casarett and Doull's Toxicology: The Basic Science of Poisons*, McGraw-Hill, Inc., Health Professions Division, 1221 Avenue of the Americas, New York, New York 10020. Fourth Edition. 1991.

Howard, P.H. et al., eds., *Handbook of Environmental Fate and Exposure Data for Organic Chemicals*, Vol. II, Lewis Publishers, Inc., 121 South Main Street, Chelsea, Michigan 48118. 1990.

Sax, N.I. and Lewis, R.J., Sr., *Dangerous Properties of Industrial Materials*, Van Nostrand Reinhold Company, 115 Fifth Avenue, New York, New York 10003. Eight Edition. 1992.

Sax, N.I. and Lewis, R.J., Sr., eds., *Hawley's Condensed Chemical Dictionary*, Van Nostrand Reinhold Company, 115 Fifth Avenue, New York, New York 10003. Eleventh Edition. 1987.

Sittig, M., *Handbook of Toxic and Hazardous Chemicals and Carcinogens*, Noyes Publications, Mill Road, Park Ridge, New Jersey 07656. 1985.

Upton, A.C. et al., *Staying Healthy in a Risky Environment*, Simon & Schuster, 1230 Avenue of the Americas, New York, New York 10020. 1993.

U.S. Environmental Protection Agency, *Integrated Risk Information System*, Office of Research and Development, Washington, D.C. 20460. February 1995.

U.S. Public Health Service (U.S. PHS), *Toxicological Profile for Toluene*, Draft Document, Agency for Toxic Substances and Disease Registry, 1600 Clifton Road, N.E. (E-28), Atlanta, Georgia 30333. May 1994.

TOXICITY PROFILE FOR DICHLOROMETHANE

The principal route of human exposure to dichloromethane (DCM) is through inhalation. DCM does not persist in the environment — it readily evaporates from soils and surface water and is eventually degraded in the atmosphere. It is slightly soluble in water and, once released to the environment, it has been reported to migrate via the air, surface water and groundwater pathways.

Deaths have been reported following acute inhalation exposure to DCM. Adverse effects on the central nervous system, liver and kidneys have been documented in laboratory animals. Impairment of central nervous system function has been demonstrated in various human studies. Cardiotoxic effects can occur in sensitive individuals (e.g., those with existing cardiac disease) when blood concentrations of carboxy-hemoglobin following DCM exposure reaches 3 - 10% (U.S. PHS, 1987). Epidemiological studies have not shown statistically significant increases in the incidence of death resulting from cancer among occupationally exposed individuals; based on animal studies, however, DCM is classified by the U.S. Environmental Protection Agency (EPA) as a Class B2, probable human carcinogen (IRIS, 1995).

BACKGROUND INFORMATION

Dichloromethane (also commonly referred to as methylene chloride) is a colorless, volatile liquid with a pleasant aromatic odor that is noticeable at 300 ppm (Sittig, 1985). It is a toxic, nonflammable organic solvent that is derived from the chlorination of methyl chloride with subsequent distillation (Sax et al., 1987).

DCM is commonly used in paint strippers and as a solvent degreasing agent to remove oils and other contaminants from the surface of metal parts in a wide range of industrial and manufacturing operations. DCM has also been used in plastics processing, as a solvent for cellulose acetate, as a blowing agent in foams, in solvent extraction operations, and as an aerosol propellant (Sax et al., 1987).

NIOSH has estimated that more than 1 million workers were exposed to DCM in 1981-83 (Howard, 1990).

PHYSICAL/CHEMICAL PROPERTIES

CAS Number: 75-09-2

Molecular Formula: CH2Cl2
Molecular Weight: 84.94
Solubility in Water: 13E+3 mg/l, 25°C
Liquid Density: 1.335 (15°/4°C)
Vapor Density: 2.9
Vapor Pressure: 434.9 mm Hg, 25°C
(Howard, 1990)
Flash Point: Nonflammable
Henry's Law Constant: 2.68E-3 atm-m3/mole
(Howard, 1990)

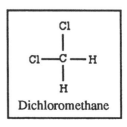

TOXICOKINETICS

Dichloromethane is readily absorbed through the respiratory and gastro-intestinal tracts and, to a lesser extent, through the skin. No information could be found on the percent of DCM absorbed by each route.

Following inhalation exposure, dichloromethane is rapidly absorbed through the lungs and into the bloodstream. Once in systemic circulation, DCM is transported (as supported by rat studies) to the liver, brain and fatty tissues; DCM has also been found in the fatty tissue of human test subjects (U.S. PHS, 1987). Human and animal (rat, mice and rabbits) studies indicate that DCM is metabolized to carbon monoxide (CO) and carbon dioxide (CO2) following inhalation exposure. Formaldehyde and formic acid, which is excreted in the urine, are two intermediate metabolic products. Elevated levels of carboxyhemoglobin in the blood of exposed individuals have been observed (U.S. PHS, 1987).

Pharmacokinetic studies point to a metabolic pathway (the GST pathway) that may be responsible for the carcinogenic effects of DCM. DCM itself is not thought to cause cancer, but rather DCM in combination with its metabolites or the action of its reactive metabolites alone has been postulated to induce carcinogenesis (U.S. PHS, 1987).

The principal route of DCM elimination is through expired air (in the form of CO, CO2 or as unchanged parent compound); formic acid, a metabolite, and a small amount of the parent compound (<2% in one study) may also be excreted in the urine (U.S. PHS, 1987). A small amount of DCM has been identified in human feces.

No human studies were identified regarding oral absorption, distribution, metabolism or excretion of DCM. Animal studies show that DCM is readily absorbed through the gastrointestinal tract and distributed to the liver, kidney, lung and fatty tissues. Following oral exposure (mice), DCM was found to be metabolized to CO and CO2. In one study, following a single oral dose, most (72%) of the DCM was eliminated through expired air with only a small fraction being excreted in the feces (<1%) or urine (2-5%) (U.S. PHS, 1987).

No studies were found regarding dermal absorption, distribution, metabolism or excretion of DCM following human exposure.

Animal studies indicate that DCM is slowly absorbed through the skin. No distribution studies following dermal absorption in laboratory animals could be found. Metabolism following dermal absorption confirms that DCM is broken down through a series of chemical reactions to CO and CO_2. No excretion studies following dermal absorption were found.

MAMMALIAN TOXICITY

Most of the animal data that exists results from inhalation and oral exposure studies. Adverse effects on the central nervous system, liver and kidneys have been documented in laboratory animals.

DCM has been tested for its genotoxicity in *in vivo* and *in vitro* test systems. DCM has been found to cause gene mutations in bacteria and yeast cells, but not in cultured mammalian cells (IRIS, 1995). Evidence of carcinogenicity in humans is inadequate. Sufficient animal data exists, however, to classify DCM as a probable human carcinogen (U.S. EPA, Class B2). Based on studies in both rats and mice, DCM was found to increase the incidence of (1) liver cancer when administered in drinking water, and (2) lung tumors, benign mammary tumors, salivary gland sarcoma and leukemia following inhalation exposure (IRIS, 1995).

Inhalation

Central nervous system (CNS) depression (including narcosis) results from acute exposure to DCM. Animal inhalation studies show that CNS effects are correlated to the concentration of DCM in the air and duration of exposure; e.g., slight narcosis @ 6,000 ppm/2.5 hrs. exposure and deep narcosis @ 16,000 to 18,000 ppm/6 hrs. exposure (U.S. PHS, 1987).

Acute and chronic exposures to DCM have been shown to alter liver structure and function in experimental animals. Adverse kidney effects resulting from inhalation exposure have also been reported. No statistically significant adverse developmental effects were found and rat studies following inhalation exposure did not reveal any adverse reproductive effects.

Ingestion/Dermal

Oral exposure studies using mice, rats and guinea pigs showed that mice and rats are equally susceptible and that guinea pigs are more susceptible to the acute/lethal effects of DCM (U.S. PHS, 1987). No animal studies regarding CNS/behavioral effects, kidney toxicity, respiratory or cardio-vascular effects following oral exposure to DCM could be found. Some data does exist, however, that DCM can produce adverse effects in the liver following acute exposure.

Subchronic exposure to DCM did not result in reproductive toxicity to male and female rats when administered in drinking water (125 ppm DCM) for 91 days (U.S. PHS, 1987). No data were found regarding toxic effects on the CNS, kidneys or cardiovascular system following subchronic oral exposure to DCM.

Chronic exposure to DCM has been shown to produce histomorphologi-cal alterations and fatty changes in the liver of mice and rats following oral exposure for 2 years (U.S. PHS, 1987). No data could be found regarding toxic effects on the CNS, kidneys or cardiovascular system following chronic exposure to DCM.

No studies regarding lethality, CNS/behavioral effects, liver/kidney toxicity, respiratory/cardiovascular development or reproductive toxicity following dermal exposure to DCM could be found.

HUMAN TOXICITY

Most of the human data that exists is derived from inhalation exposure studies. Deaths have been reported following acute inhalation exposure to DCM. Impaired central nervous system function is considered to be the primary toxic effect following short-term inhalation exposure. Cardiotoxic effects can occur in sensitive individuals (e.g., those with existing cardiac disease) when blood concentrations of carboxyhemoglobin reaches 3 - 10% (U.S. PHS, 1987). Epidemiologic studies have not shown statistically significant increases in the incidence of death resulting from cancer among occupationally exposed individuals; based on animal studies, however, DCM is classified by the U.S. EPA, and therefore should be regarded, as a probable human carcinogen.

Inhalation

The principal route of human exposure to DCM is through inhalation. Two documented cases of human death following acute inhalation expo-sure to methylene chloride during paint-stripping operations have been reported (U.S. PHS, 1987). Liver and CNS effects have been observed in humans following inhalation exposure.

Impairment of CNS function is the primary adverse effect resulting from short-term exposure to DCM; long-term exposure, however, also produces CNS effects — i.e., neurotoxicity specifically has been linked to occupa-tional exposure (@ concentrations up to 100 ppm for 6 months to 2 years) where the range of symptoms included headaches, dizziness, nausea, memory loss, tingling in hands and feet and loss of consciousness (U.S. PHS, 1987).

Though limited human data exist, case reports do suggest that DCM alters liver function in humans following inhalation exposure. Epidemiologic studies failed to show an association between DCM exposure and adverse kidney effects. Cardiotoxic effects can occur in sensitive individuals (e.g., those with existing cardiac disease) when blood concentrations of carboxyhemoglobin reaches 3 - 10% (U.S. PHS, 1987). DCM has been shown to cause respiratory irritation in humans. No developmental or reproductive toxicity data following inhalation exposure could be found.

Ingestion/Dermal

No human studies regarding lethality, CNS/behavioral effects, liver/kidney toxicity, cardiovascular effects, developmental or reproductive toxicity following oral exposure to DCM could be found.

No human studies regarding lethality, CNS/behavioral effects, liver/kidney toxicity, cardiovascular effects, developmental or reproductive toxicity following dermal exposure to DCM could be found.

ECOLOGICAL TOXICITY

Dichloromethane has been reported to biodegrade completely when introduced to acclimated microbial populations under aerobic conditions; no information was found regarding biodegradation in natural aquatic systems or groundwater (Howard, 1990). High concentrations may be toxic to microorganisms and, at least initially, may interfere with naturally occurring biodegradation processes. Based on calculated values, DCM is not expected to bioaccumulate or bioconcentrate in aquatic organisms to a significant degree (U.S. PHS, 1987), though it has been found at low concentrations in the tissues of bottom fish (0.7 ppm), oysters (7.8 ppb), and clams (4.5 ppb) (Howard, 1990).

Dichloromethane's high vapor pressure, low water solubility, low bioconcentration factor and propensity toward atmospheric degradation suggests that significant levels of DCM would not persist long enough in the environment to present a significant long-term ecological threat. Large releases of DCM to a waterway — resulting from a spill or accidental discharge for example — could, however, be acutely toxic to aquatic life and result in a measurable disturbance of the receiving ecosystem. Representative marine and freshwater species appear to be equally sensitive to the acute effects of DCM (see Regulations and Standards).

No data on domestic or wild animals could be found.

FATE AND TRANSPORT

Though some evidence does exist that DCM may be formed by natural processes, these sources are not thought "to contribute significantly to global releases" (U.S. PHS, 1987). Dichloromethane has been found in drinking water (surface and groundwater supplies), industrial effluents, surface waters (both fresh and seawater), sediment/soil samples and in the atmosphere at remote, rural, urban and suburban locations.

Once released to the environment, dichloromethane quickly volatilizes and is eventually degraded in the atmosphere. Atmospheric degradation is principally by reaction with photochemically produced hydroxyl radicals with half-lives on the order of several months (Howard, 1990). Small amounts (~1%) of dichloromethane reach the stratosphere where it degrades by reaction with chlorine radicals and by photolysis (U.S. PHS, 1987). Due to its water solubility (13,000 mg/l @25°C), dichloromethane that is not degraded in the troposphere will be washed out with rain and reintroduced to the hydrosphere.

Dichloromethane readily evaporates from surface water, the rate being dependent upon temperature and mixing. Though biodegredation is possible with controlled, acclimated microbial populations, it is thought to be a very slow fate process in natural surface water environments when compared to evaporation (Howard, 1990). Hydrolysis in water is slow (estimated half-life of 18 months @ 25°C) and is dependent upon pH and temperature (U.S. PHS, 1987). Calculated bioconcentration and soil partition coefficients suggest that dichloromethane would not be expected to bioconcentrate in aquatic organisms or adsorb to low organic content soils to any significant degree.

Dichloromethane releases to soil are expected to evaporate rather quickly, though leaching to groundwater is possible and can play an important role in the transport and distribution of dichloromethane in the environment. Though the fate of dichloromethane in groundwater is unknown, hydrolysis is not thought to be an important degradation process (Howard, 1990).

REGULATIONS AND STANDARDS (U.S. PHS, 1987)

Regulations
Occupational Safety and Health Administration (OSHA)
OSHA Permissible Exposure Limit: TWA 500 ppm; Ceiling 1000 ppm; Max. Peak 2000 ppm

Resource Conservation and Recovery Act
U.S. EPA Listed Hazardous Waste:

F001 Spent halogenated solvent used in degreasing
F002 Spent halogenated solvent
U080 Toxic waste

CERCLA/Superfund
Reportable Quantity for Accidental Release: 1000 lbs.

Guidelines
Worker Exposure Guidelines
ACGIH TLV: TWA 50 ppm
NIOSH: TWA 75 ppm; CL 500 ppm/15 minute avg.

Safe Drinking Water Act
Maximum Contaminant Level Goal: 0 mg/l
Maximum Contaminant Level: 0.005 mg/l

Drinking Water Health Advisory
1 day (child): 1.33E+1 mg/l
10 (child): 1.5E+0 mg/l

Clean Water Act
Ambient Water Quality Criteria for Human Health
Water & Fish: 1.9E-1 ug/l
Fish Only: 1.57E+1 ug/l

Ambient Water Quality Criteria for Aquatic Organisms
Freshwater: 1.1E+4 ug/l, acute toxicity basis (LEC)
Marine:1.2E+4 ug/l, acute toxicity basis (LEC)
6.4E+3 ug/l, chronic toxicity basis (LEC)

REFERENCES (Dichloromethane)

Howard, P.H. et al., eds., *Handbook of Environmental Fate and Exposure Data for Organic Chemicals,* Vol. II, Lewis Publishers, Inc., 121 South Main Street, Chelsea, Michigan 48118. 1990.

Sax, N.I. and R.J. Lewis, Sr., *Dangerous Properties of Industrial Materials,* Van Nostrand Reinhold Company, 115 Fifth Avenue, New York, New York 10003. Eight Edition. 1992.

Sax, N.I. and R.J. Lewis, Sr., eds., *Hawley's Condensed Chemical Dictionary,* Van Nostrand Reinhold Company, 115 Fifth Avenue, New York, New York 10003. Eleventh Edition. 1987.

Sittig, M., *Handbook of Toxic and Hazardous Chemicals and Carcinogens,* Noyes Publications, Mill Road, Park Ridge, New Jersey 07656. 1985.

REFERENCES (Continued)

U.S. Environmental Protection Agency, *Integrated Risk Information System,* Office of Research and Development, Washington, D.C. 20460. February 1995.

U.S. Public Health Service (U.S. PHS), *Toxicological Profile for Methylene Chloride,* Draft Document, Agency for Toxic Substances and Disease Registry, 1600 Clifton Road, N.E. (E-28), Atlanta, Georgia 30333. December 1987.

Appendix D

Test Methods for Evaluating Solid Waste
SW-846, Sampling Plan

"Test Methods for Evaluating Solid Waste" is a two volume set containing thirteen chapters. Most of the text presented specifies test methods for analytes/properties of concern and laboratory methods for waste determinations. Chapter 9, entitled "Sampling Plan", however, provides guidance to generators and others on the design and implementation of a scientifically defensible solid and hazardous waste sampling plan. For the convenience of readers, key text from Chapter 9 is reprinted below. The complete text of U.S. EPA guidance on representative sampling procedures and accepted laboratory test methods can be found in "Test Methods for Evaluating Solid Waste, Physical/Chemical Methods", SW-846. This set can be purchased from the U.S. Department of Commerce, National Technical Information Service, Springfield, VA 22161. FY 1995 purchase price for a hard copy of the 3rd edition, including final updates, is $358; the complete text is also available from NTIS on CD ROM for $275. Finally, SW-846 may be available for review, free of charge, at your state environmental protection agency, commercial hazardous waste testing laboratory, university or U.S. EPA regional library.

DESIGN AND DEVELOPMENT

The initial — and perhaps most critical — element in a program designed to evaluate the physical and chemical properties of a solid waste is the plan for sampling the waste. It is understandable that analytical studies, with their sophisticated instrumentation and high cost, are often perceived as the dominant element in a waste characterization program. Yet, despite that sophistication and high cost, analytical data generated by a scientifically defective sampling plan have limited utility, particularly in the case of regulatory proceedings.

This section of the manual addresses the development and implementation of a scientifically credible sampling plan for a solid waste and the documentation of the chain of custody for such a plan. The information presented in this section is relevant to the sampling of any solid waste which has been defined by the EPA in its regulations for the identification and listing of hazardous wastes to include solid, semisolid, liquid, and contained gaseous materials. However, the physical and chemical diversity of those materials, as well as the dissimilarity of storage facilities (lagoons, open piles, tanks, drums, etc.) and sampling equipment associated with them, preclude a detailed consideration of any specific sampling plan. Consequently, because the burden of responsibility for developing a technically sound sampling plan rests with the waste producer, it is advisable that he/she seek competent advice before designing a plan. This is particularly true in the early developmental stages of a sampling plan, at which time at least a basic understanding of applied statistics is required. Applied statistics is the science of employing techniques that allow the uncertainty of inductive inferences (general conclusions based on partial knowledge) to be evaluated.

DEVELOPMENT OF APPROPRIATE SAMPLING PLANS

An appropriate sampling plan for a solid waste must be responsive to both regulatory and scientific objectives. Once those objectives have been clearly identified, a suitable sampling strategy, predicated upon fundamental statistical concepts, can be developed.

Regulatory and Scientific Objectives

The EPA, in its hazardous waste management system, has required that certain solid wastes be analyzed for physical and chemical properties. It is mostly chemical properties that are of concern, and, in the case of a number of chemical contaminants, the EPA has promulgated levels (regulatory thresholds) that cannot be equaled or exceeded. The regulations pertaining to the management of hazardous wastes contain three references regarding the sampling of solid wastes for analytical properties. The first reference, which occurs throughout the regulations, requires that representative samples of waste be collected and defines representative samples as exhibiting average properties of the whole waste. The second reference, which pertains just to petitions to exclude wastes from being listed as hazardous wastes, specifies that enough samples (but in no case less that four samples) be collected over a period of time sufficient to represent the variability of wastes. The third reference, which applies only to ground water monitoring systems, mandates that four replicates (subsamples) be taken from each ground water sample intended for chemical analysis and that the mean concentration and variance for each chemical constituent be calculated from those four subsamples and compared with background levels for ground water. Even the statistical test to be employed in that comparison is specified (Student's t-test).

The first of the above-described references addresses the issue of sampling accuracy, and the second and third references focus on sampling variability or, conversely, sampling precision (actually the third reference relates to analytical variability, which in many statistical tests, is indistinguishable from true sampling variability). Sampling accuracy (the closeness of a sample value to its true value) and sampling precision (the closeness of repeated sample values) are also the issues of overriding importance in any scientific assessment of sampling practices. Thus, from both regulatory and scientific perspectives, the primary objectives of a sampling plan for a solid waste are twofold: namely, to collect samples that will allow measurements of the chemical properties of the waste that are both accurate and precise. If the chemical measurements are sufficiently accurate and precise, they will be considered reliable estimates of the chemical properties of the waste.

It is now apparent that a judgment must be made as to the degree of sampling accuracy and precision that is required to estimate reliably the chemical characteristics of a solid waste for the purpose of comparing those characteristics with applicable regulatory thresholds. Generally, high accuracy and high precision are required if one or more chemical contaminants of a solid waste are present at a concentration that is close to the applicable regulatory threshold. Alternatively, relatively low accuracy and low precision can be tolerated if the contaminants of concern occur at levels far below or far above their applicable thresholds. However, a word of caution is in order. Low sampling precision is often associated with considerable savings in analytical, as well as sampling, costs and is clearly recognizable even in the simplest of statistical tests. On the other hand, low sampling accuracy may not entail cost savings and is always obscured in statistical tests (i.e., it cannot be evaluated). Therefore, although it is desirable to design sampling plans for solid wastes to achieve only the minimally required precision (at least two samples of material are required for any estimate of precision), it is prudent to design the plans to attain the greatest possible accuracy.

Fundamental Statistical Concepts

Statistical techniques for obtaining accurate and precise samples are relatively simple and easy to implement. Sampling accuracy is usually achieved by some form of random sampling. In random sampling, every unit in the population (e.g., every location in a lagoon used to store a solid waste) has a theoretically equal chance of being sampled and measured. Consequently, statistics generated by the sample (e.g., the mean and, to a lesser degree, the standard error) are unbiased (accurate) estimators of true population parameters (e.g., the confidence interval for the population mean). In other words, the sample is representative of the population. One of the commonest methods of selecting a random sample is to divide the population by an imaginary grid, assign a series of consecutive numbers to the units of the grid, and select the numbers (units) to be sampled through the use of random-numbers table (such a table can be found in any text on basic statistics). It is important to emphasize that a haphazardly selected sample is not a suitable substitute for a randomly selected sample. That is because there is no assurance that a person performing undisciplined sampling will not consciously or subconsciously favor the selection of certain units of the population, thus causing the sample to be unrepresenta-tive of the population.

Sampling precision is most commonly achieved by taking an appropri-ate number of samples from the population. Another technique for increasing sampling precision is to maximize the physical size (weight or volume) of the samples that are collected. That has the effect of minimizing between-sample variation and, consequently, decreasing the standard error. Increasing the number or size of samples taken from a population, in addition to increasing sampling precision, has the secondary effect of increasing sampling accuracy.

In summary, reliable information concerning the chemical properties of a solid waste is needed for the purpose of comparing those properties with applicable regulatory thresholds. If chemical information is to be consid-ered reliable, it must be accurate and sufficiently precise. Accuracy is usually achieved by incorporating some form of randomness into the selection process for the samples that generate the chemical information. Sufficient precision is most often obtained by selecting an appropriate number of samples.

There are a few ramifications of the above-described concepts that merit elaboration. If, for example, as in the case of semiconductor etching solutions, each batch of a waste is completely homogeneous with regard to the chemical properties of concern and that chemical homogeneity is constant (uniform) over time (from batch to batch), a single sample collected from the waste at an arbitrary location and time would theoreti-cally generate an accurate and precise estimate of the chemical properties. However, most wastes are heterogeneous in terms of their chemical properties. If a batch of waste is randomly heterogeneous with regard to its chemical characteristics and that random chemical heterogeneity remains constant from batch to batch, accuracy and appropriate precision can usually be achieved by simple random sampling. In that type of sampling, all units in the population (essentially all locations or points in all batches of waste from which a sample could be collected) are identified, and a suitable number of samples is randomly selected from the population. More complex stratified random sampling is appropriate if a batch of waste is known to be nonrandomly heterogeneous in terms of its chemical properties and/or nonrandom chemical heterogeneity is known to exist from batch to batch. In such cases, the population is stratified to isolate the known sources of nonrandom chemical heterogeneity.

After stratification, which may occur over space (locations or points in a batch of waste) and/or time (each batch of waste), the units in each stratum are numerically identified, and a simple random sample is taken from each stratum. As previously intimated, both simple and stratified random sampling generate accurate estimates of the chemical properties of a solid waste. The advantage of stratified random sampling over simple random sampling is that, for a given number of samples and a given sample size, the former technique often results in a more precise estimate of chemical properties of a waste (a lower value of standard error) than the latter technique. However, greater precision is likely to be realized only if a waste exhibits substantial nonrandom chemical heterogeneity and stratification efficiently "divides" the waste into strata that exhibit maximum between-strata variability and minimum within-strata variability. If that does not occur, stratified random sampling can produce results that are less precise than in the case of simple random sampling. Therefore, it is reasonable to select stratified random sampling over simple random sampling only if the distribution of chemical contaminants in a waste is sufficiently known to allow an intelligent identification of strata and at least two or three samples can be collected in each stratum. If a strategy employing stratified random sampling is selected, a decision must be made regarding the allocation of sampling effort among strata. When chemical variation within each stratum can be estimated with a great degree of detail, samples should be optimally allocated among strata, i.e., the number of samples collected from each stratum should be directly proportional to the chemical variation encountered in the stratum. When detailed information concerning chemical variability within strata is not available, samples should be proportionally allocated among strata, i.e., sampling effort in each stratum should be directly proportional to the size of the stratum.

Simple random sampling and stratified random sampling are types of probability sampling, which, because of a reliance upon mathematical and statistical theories, allows an evaluation of the effectiveness of sampling procedures. Another type of probability sampling is systematic random sampling, in which the first unit to be collected from a population is randomly selected, but all subsequent units are taken at fixed space or time intervals. An example of systematic random sampling is the sampling of a waste lagoon along a transect in which the first sampling point on the transect is 1 m from a randomly selected location on the shore and subsequent sampling points are located at 2-m intervals along the transect. The advantages of systematic random sampling over simple random sampling and stratified random sampling are the ease with which samples are identified and collected (the selection of the first sampling unit determines the remainder of the units) and, sometimes an increase in precision. In certain cases, for example, systematic random sampling might be expected to be a little more precise than stratified random sampling with one unit per stratum because samples are distributed more evenly over the population. As will be demonstrated shortly, disadvantages of systematic random sampling are the poor accuracy and precision that can occur when unrecognized trends or cycles occur in the population. For those reasons, systematic random sampling is recommended only when a population is essentially random or contains at most a modest stratification. In such cases, systematic random sampling would be employed for the sake of convenience, with little expectation of an increase in precision over other random sampling techniques.

Probability sampling is contrasted with authoritative sampling, in which an individual who is well acquainted with the solid waste to be sampled selects a sample without regard to randomization. The validity of data gathered in that manner is totally dependent on the knowledge of the sampler and, although valid data can sometimes be obtained, authoritative sampling is not recommended for the chemical characterization of most wastes.

It may now be useful to offer a generalization regarding the four sampling strategies that have been identified for solid wastes. If little or no information is available concerning the distribution of chemical contaminants of a waste, simple random sampling is the most appropriate sampling strategy. As more information is accumulated for the contaminants of concern, greater consideration can be given (in order of the additional information required) to stratified random sampling, systematic random sampling, and perhaps, authoritative sampling.

The validity of a Confidence Interval (CI) for the true mean (μ) concentration of a chemical contaminant of a solid waste is, as previously noted, based on the assumption that individual concentrations of the contaminant exhibit a normal distribution. This is true regardless of the strategy that is employed to sample the waste. Although there are computational procedures for evaluating the correctness of the assumption of normality, those procedures are meaningful only if a large number of samples are collected from a waste. Because sampling plans for most solid wastes entail just a few samples, one can do little more than superficially examine resulting data for obvious departures from normality (this can be done by simple graphical methods), keeping in mind that even if individual measurements of a chemical contaminant of a waste exhibit a considerable abnormal distribution, such abnormality is not likely to be the case for sample means, which are our primary concern. One can also compare the mean of the sample (X) with the variance of the sample (s^2). In a normally distributed population, X would be expected to be greater than s^2 (assuming that the number of samples [n] is reasonable large). If that is not the case, the chemical contaminant of concern may be characterized by a Poisson distribution (X is approximately equal to s^2) or a negative binomial distribution (x is less than s^2). In the former circumstance, normality can often be achieved by transforming data according to the square root transformation. In the latter circumstance, normality may be realized through use of the arcsine transformation. If either transformation is required, all subsequent statistical evaluations must be performed on the transformed scale.

Finally, it is necessary to address the appropriate number of samples to be employed in the chemical characterization of a solid waste. As has already been emphasized, the appropriate number of samples is the least number of samples required to generate a sufficiently precise estimate of the true mean concentration of a chemical contaminant of a waste. From the perspective of most waste producers, that means the minimal number of samples needed to demonstrate that the upper limit of the CI for μ is less than the applicable regulatory threshold. (Author's Note: Formulas and procedures for "estimating [an] appropriate sampling effort" can be found in the complete text of SW-846. Examples that illustrate the use of simple random sampling, stratified random sampling, systematic random sampling, composite sampling, and subsampling are also presented in SW-846.)

IMPLEMENTATION

This section discusses the implementation of a sampling plan for the collection of a "solid waste," as defined by Section 261.2 of the Resource Conservation and Recovery Act regulations. Due to the uniqueness of each sampling effort, the following discussion is in the general form of guidance which, when applied to each sampling effort, should improve and document the quality of the sampling and the representativeness of samples.

The following subsections address elements of a sampling effort in a logical order, from defining objectives through compositing samples prior to analysis.

DEFINITION OF OBJECTIVES

After verifying the need for sampling, those personnel directing the sampling effort should define the program's objectives. The need for a sampling effort should not be confused with the objective. When management, a regulation, or a regulatory agency requires sampling, the need for sampling is established but the objectives must be defined.

The primary objective of any waste sampling effort is to obtain information that can be used to evaluate a waste. It is essential that the specific information needed and its uses are defined in detail at this stage. The information needed is usually more complex than just a concentration of a specified parameter; it may be further qualified (e.g., by sampling location or sampling time.) The manner in which the information is to be used can also have a substantial impact on the design of a sampling plan. (Are the data to be used in a qualitative manner? If quantitative, what are the accuracy and precision requirements?)

All pertinent information should be gathered. For example, if the primary objective has been roughly defined as "collecting samples of waste which will be analyzed to comply with environmental regulations," then ask the following questions:

1. The sampling is being done to comply with which environmental regulation? Certain regulations detail specific or minimum protocols (e.g., exclusion petitions as defined in §260.22 or RCRA regulations); the sampling effort must comply with these regulatory requirements.

2. The collected samples are to be analyzed for which parameters? Why those and not others? Should the samples be analyzed for more or fewer parameters?

3. What waste is to be sampled: the waste as generated? the waste prior to or after mixing with other wastes or stabilizing agents? the waste after aging or drying or just prior to disposal?   Should waste disposed of 10 years ago be sampled to acquire historical data?

4. What is the end-use of the generated data base? What are the required degrees of accuracy and precision?

By asking such questions, both the primary objective and specific sampling, analytical, and data objectives can be established.

SAMPLING PLAN CONSIDERATIONS

The sampling plan is usually a written document that describes the objectives and details the individual tasks of a sampling effort and how they will be performed. (Under unusual circumstances, time may not allow for the sampling plan to be documented in writing, e.g., sampling during an emergency spill. When operating under these conditions, it is essential that the person directing the sampling effort be aware of the various elements of a sampling plan.) The more detailed the sampling plan, the less the opportunity for oversight or misunderstanding during sampling, analysis, and data treatment.

To ensure that the sampling plan is designed properly, it is wise to have all aspects of the effort represented. Those designing the sampling plan should include the following personnel:

1.   An end-user of the data, who will be using the data to attain program

objectives and thus would be best prepared to ensure that the data objectives are understood and incorporated into the sampling plan.

2. An experienced member of the field team who will actually collect samples, who can offer hands-on insight into potential problems and solutions, and who, having acquired a comprehensive understanding of the entire sampling effort during the design phase, will be better prepared to implement the sampling plan.

3. An analytical chemist, because the analytical requirements for sampling, preservation, and holding times will be factors around which the sampling plan will be written. A sampling effort cannot succeed if an improperly collected or preserved sample or an inadequate volume of sample is submitted to the laboratory for chemical, physical, or biological testing. The appropriate analytical chemist should be consulted on these matters.

4. An engineer should be involved if a complex manufacturing process is being sampled. Representation of the appropriate engineering discipline will allow for the optimization of sampling locations and safety during sampling and should ensure that all waste-stream variations are accounted for.

5. A statistician, who will review the sampling approach and verify that the resulting data will be suitable for any required statistical calculations or decisions.

6. A quality assurance representative, who will review the applicability of standard operating procedures and determine the number of blanks, duplicates, spike samples, and other steps required to document the accuracy and precision of the resulting data base.

Waste

The sampling plan must address a number of factors in addition to statistical considerations. Obviously, one of the most important factors is the waste itself and its properties. The following waste properties are examples of what must be considered when designing a sampling plan:

1. Physical state: The physical state of the waste will affect most aspects of a sampling effort. The sampling device will vary according to whether the sample is liquid, gas, solid, or multiphasic. It will also vary according to whether the liquid is viscous or free-flowing, or whether the solid is hard or soft, powdery, monolithic, or claylike.

Wide-mouth sample containers will be needed for most solid samples and for sludges or liquids with substantial amounts of suspended matter. Narrow-mouth containers can be used for other wastes, and bottles with air-tight closures will be needed for gas samples or gases adsorbed on solids or dissolved in liquids.

The physical state will also affect how sampling devices are deployed. A different plan will be developed for sampling a soil-like waste that can easily support the weight of a sampling team and its equipment than for a lagoon filled with a viscous sludge or a liquid waste.

The sampling strategy will have to vary if the physical state of the waste allows for stratification (e.g., liquid wastes that vary in density or viscosity or have a suspended solid phase), homogenization or random heterogeneity.

2. Volume: The volume of the waste, which has to be represented by the samples collected, will have an effect upon the choice of sampling equipment and strategies. Sampling a 40-acre lagoon requires a different approach from sampling a 4-sq-ft container. Although a 3-ft depth can be sampled with a Coliwasa or a drum thief, a weighted bottle may be required to sample a 50-ft depth.

3. Hazardous properties: Safety and health precautions and methods of sampling and shipping will vary dramatically with the toxicity, ignitability, corrosivity, and reactivity of the waste.

4. Composition: The chosen sampling strategy will reflect the homogeneity, random heterogeneity, or stratification of the waste in time or over space.

Site

Site-specific factors must be considered when designing a sampling plan. A thorough examination of these factors will minimize oversights that can affect the success of sampling and prevent attainment of the program objectives. At least one person involved in the design and implementation of the sampling plan should be familiar with the site, or a presampling site visit should be arranged. If nobody is familiar with the site and a visit cannot be arranged, the sampling plan must be written to account for the possible contingencies. Examples of site-specific factors that should be considered follow:

1. Accessibility: The accessibility of waste can vary substantially. Some wastes are accessed by the simple turning of a valve; others may require that an entire tank be emptied or that heavy equipment be employed. The accessibility of a waste at the chosen sampling location must be determined prior to design of a sampling plan.

2. Waste generation and handling: The waste generation and handling process must be understood to ensure that collected samples are representative of the waste. Factors which must be known and accounted for in the sampling plan include: if the waste is generated in batches; if there is a change in the raw materials used in a manufacturing process; if waste composition can vary substantially as a function of process temperatures or pressures; and if storage time after generation may vary.

3. Transitory events: Start-up, shut-down, and maintenance transients can result in the generation of a waste that is not representative of the normal waste stream. If a sample was unknowingly collected at one of these intervals, incorrect conclusions could be drawn.

4. Climate: The sampling plan should specify any clothing needed for personnel to accommodate any extreme heat or cold that may be encountered. Dehydration and extensive exposure to sun, insects, or poisonous snakes must be considered.

5. Hazards: Each site can have hazards — both expected and unexpected. For example, a general understanding of a process may lead a simple team to be prepared for dealing with toxic or reactive material, but not for dealing with an electrical hazard or the potential for suffocation in a confined space. A thorough sampling plan will include a health and safety plan that will counsel team members to be alert to potential hazards.

Equipment

The choice of sampling equipment and sample containers will depend upon the previously described waste and site considerations. For the following reasons, the analytical chemist will play an important role in the selection of sampling equipment:

1. The analytical chemist is aware of the potential interactions between sampling equipment or container material with analytes of interest. As a result, he/she can suggest a material that minimizes

losses by adsorption, volatilization, or contamination caused by leaching from containers or sampling devices.

2. The analytical chemist can specify cleaning procedures for sampling devices and containers that minimize sample contamination and cross contamination between consecutive samples.

3. The analytical chemist's awareness of analyte-specific properties is useful in selecting the optimum equipment (e.g., choice of sampling devices that minimize agitation for those samples that will be subjected to analysis for volatile compounds).

The final choice of containers and sampling devices will be made jointly by the analytical chemist and the group designing the sampling plan. The factors that will be considered when choosing a sampling device are:

1. Negative contamination: The potential for the measured analyte concentration to be artificially low because of losses from volatilization or adsorption.

2. Positive contamination: The potential for the measured analyte to be artificially high because of leaching or the introduction of foreign matter into the sample by particle fallout or gaseous air contaminants.

3. Cross contamination: A type of positive contamination caused by the introduction of part of one sample into a second sample during sampling, shipping, or storage.

4. Required Sample Volume for physical and/or chemical analysis.

5. "Ease of use" of the sampling device and containers under the conditions that will be encountered on site. This includes the ease of shipping to and from the site, ease of deployment, and ease of cleaning.

6. The degree of hazard associated with the deployment of one sampling device versus another.

7. Cost of the sampling device and of the labor for its deployment.

This section describes examples of sampling equipment and suggests potential uses for this equipment. Some of these devices are commercially available, but others will have to be fabricated by the user. The information in this section is general in nature and therefore limited.

Because each sampling situation is unique, the cited equipment and applications may have to be modified to ensure that a representative sample is collected and its physical and chemical integrity are maintained. It is the responsibility of those persons conducting sampling programs to make the appropriate modifications.

Table 14 (Chapter 4) contains examples of sampling equipment and potential applications. It should be noted that these suggested sampling devices may not be applicable to a user's situation due to waste- or site-specific factors. For example, if a waste is highly viscous or if a solid is clay-like, these properties may preclude the use of certain sampling devices. The size and depth of a lagoon or tank, or difficulties associated with accessing the waste, may also preclude use of a given device or require modification of its deployment.

The most important factors to consider when choosing containers for hazardous waste samples are compatibility with the waste, cost, resistance to breakage, and volume. Containers must not distort, rupture, or leak as a result of chemical reactions with constituents of waste samples. Thus, it is important to have some idea of the properties and composition of the waste. The containers must have adequate wall thickness to withstand handling during sample collection and transport to the laboratory. Containers with wide mouths are often desirable to facilitate transfer of samples from samplers to containers. Also, the containers must be large enough to contain the optimum sample volume.

Containers for collecting and storing hazardous waste samples are usually made of plastic or glass. Plastics that are commonly used to make the containers include high-density or linear polyethylene (LPE), conventional polyethylene, polypropylene, polycarbonate, Teflon FEP (fluorinated ethylene propylene), polyvinyl chloride (PVC), or polymethylpentene. Teflon FEP is almost universally usable due to its chemical inertness and resistance to breakage. However, its high cost severely limits its use. LPE, on the other hand, usually offers the best combination of chemical resistance and low cost when samples are to be analyzed for inorganic parameters.

Glass containers are relatively inert to most chemicals and can be used to collect and store almost all hazardous waste samples, except those that contain strong alkali and hydrofluoric acid. Glass soda bottles are suggested due to their low cost and ready availability. Borosilicate glass containers, such as Pyrex and Corex, are more inert and more resistant to breakage than soda glass, but are expensive and not always readily available. Glass containers are generally more fragile and much heavier than plastic containers. Glass or FEP containers must be used for waste samples that will be analyzed for organic compounds.

The containers must have tight, screw-type lids. Plastic bottles are usually provided with screw caps made of the same material as the bottles. Buttress threads are recommended. Cap liners are not usually required for plastic containers. Teflon cap liners should be used with glass containers supplied with rigid plastic screw caps. (These caps are usually provided with waxed paper liners.) Teflon liners may be purchased from plastic specialty supply houses (e.g., Scientific Specialties Service, Inc., P.O. Box 352, Randallstown, Maryland 21133). Other liners that may be suitable are polyethylene, polypropylene, and neoprene plastics.

If the samples are to be submitted for analysis of volatile compounds, the samples must be sealed in air-tight containers.

Prior to sampling, a detailed equipment list should be compiled. This equipment list should be comprehensive and leave nothing to memory. The categories of materials that should be considered are:

1. Personnel equipment, which will include boots, rain gear, disposable coveralls, face masks and cartridges, gloves, etc.

2. Safety equipment, such as portable eyewash stations and a first-aid kit.

3. Field test equipment, such as pH meters and Draeger tube samplers.

4. An ample supply of containers to address the fact that once in the field, the sampling team may want to collect 50% more samples than originally planned or to collect a liquid sample, although the sampling plan had specified solids only.

5. Additional sampling equipment for use if a problem arises, e.g., a tool kit.

6. Shipping and office supplies, such as tape, labels, shipping forms, chain-of-custody forms and seals, field notebooks, random-number tables, scissors, pens, etc.

Composite Liquid Waste Sampler (Coliwasa)

The Coliwasa is a device employed to sample free-flowing liquids and slurries contained in drums, shallow tanks, pits, and similar containers. It is especially useful for sampling wastes that consist of several immiscible liquid phases.

The coliwasa consists of a glass, plastic, or metal tube equipped with an end closure that can be opened and closed while the tube is submerged in the material to be sampled.

Weighted Bottle

This sampler consists of a glass or plastic bottle, sinker, stopper, and a line that is used to lower, raise, and open the bottle. The weighted bottle samples liquids and free-flowing slurries. A weighted bottle with line is built to the specifications in ASTM Methods D270 and E300.

Dipper

The dipper consists of a glass or plastic beaker clamped to the end of a two- or three-piece telescoping aluminum or fiberglass pole that serves as the handle. A dipper samples liquids and free-flowing slurries. Dippers are not available commercially and must be fabricated.

Thief

A thief consists of two slotted concentric tubes, usually made of stainless steel or brass. The outer tube has a conical pointed tip that permits the sampler to penetrate the material being sampled. The inner tube is rotated to open and close the sampler. A thief is used to sample dry granules or powdered wastes whose particle diameter is less than one-third the width of the slots. A thief is available at laboratory supply stores.

Trier

A trier consists of a tube cut in half lengthwise with a sharpened tip that allows the sampler to cut into sticky solids and to loosen soil. A trier samples moist or sticky solids with a particle diameter less than one-half the diameter of the trier. Triers 61 to 100 cm long and 1.27 to 2.54 cm in diameter are available at laboratory supply stores. A large trier can be fabricated.

Auger

An auger consists of sharpened spiral blades attached to a hard metal central shaft. An auger samples hard or packed solid wastes or soil. Augers are available at hardware and laboratory supply stores.

Scoops and Shovels

Scoops and shovels are used to sample granular or powdered material in bins, shallow containers, and conveyor belts. Scoops are available at laboratory supply houses. Flat-nosed shovels are available at hardware stores.

Bailer

The bailer is employed for sampling well water. It consists of a container attached to a cable that is lowered into the well to retrieve a sample. Bailers can be of various designs. The simplest is a weighted bottle or basally capped length of pipe that fills from the top as it is lowered into the well. Some bailers have a check valve, located at the base, which allows water to enter from the bottom as it is lowered into the well. When the bailer is lifted, the check valve closes, allowing water in the bailer to be brought to the surface. More sophisticated bailers are available that remain open at both ends while being lowered, but can be sealed at both top and bottom by activating a triggering mechanism from the surface. This allows more reliable sampling at discrete depths within a well. Perhaps the best-known bailer of this latter design is the Kemmerer sampler.

Bailers generally provide an excellent means for collecting samples from monitoring wells. They can be constructed from a wide variety of materials compatible with the parameter of interest. Because they are relatively inexpensive, bailers can be easily dedicated to an individual well to minimize cross contamination during sampling. If not dedicated to a well, they can be easily cleaned to prevent cross contamination. Unfortunately, bailers are frequently not suited for well evacuation because of their small volume.

Quality Assurance and Quality Control

Quality assurance (QA) can briefly be defined as the process for ensuring that all data and the decisions based on these data are technically sound, statistically valid, and properly documented. Quality control (QC) procedures are the tools employed to measure the degree to which these quality assurance objectives are met.

A data base cannot be properly evaluated for accuracy and precision unless it is accompanied by quality assurance data. In the case of waste evaluation, these quality assurance data result from the implementation of quality control procedures during sampling and analysis. Quality control requirements for specific analytical methods are given in detail in each method in this manual; in this subsection, quality assurance and quality control procedures for sampling will be discussed.

Quality control procedures that are employed to document the accuracy and precision of sampling are:

1. Trip Blanks: Trip blanks should accompany sample containers to and from the field. These samples can be used to detect any contamination or cross-contamination during handling and transportation.

2. Field Blanks: Field blanks should be collected at specified frequencies, which will vary according to the probability of contamination or cross-contamination. Field blanks are often metal- and/or organic-free water aliquots that contact sampling equipment under

field conditions and are analyzed to detect any contamination from sampling equipment, cross contamination from previously collected samples, or contamination from conditions during sampling (e.g., airborne contaminants that are not from the waste being sampled).

3. Field Duplicates: Field duplicates are collected at specified frequencies and are employed to document precision. The precision resulting from field duplicates is a function of the variance of waste composition, the variance of the sampling technique, and the variance of the analytical technique.

4. Field Spikes: Field spikes are infrequently used to determine the loss of parameters of interest during sampling and shipment to the laboratories. Because spiking is done in the field, the making of spiked samples or spiked blanks is susceptible to error. In addition, compounds can be lost during spiking, and equipment can be contaminated with spiking solutions. To eliminate these and other problems, some analysts spike blanks or matrices similar to the waste in the laboratory and ship them, along with sample containers, to the field. This approach also has its limitation because the matrix and the handling of the spike are different from those of the actual sample. In all cases, the meaning of a low field-spike recovery is difficult to interpret, and thus, field spikes are not commonly used.

In addition to the above quality control samples, a complete quality assurance program will ensure that standard operating procedures (SOPs) exist for all essential aspects of a sampling effort. SOPs should exist for the following steps in a sampling effort:

1. Definition of objectives.
2. Design of sampling plans.
3. Preparation of containers and equipment (refer to the specific analytical methods).
4. Maintenance, calibration, and cleaning of field equipment (refer to instrument manuals or consult a chemist for cleaning protocols).
5. Sample preservation, packaging, and shipping.
6. Health and safety protocols.
7. Chain-of-custody protocols.

In addition to the above protocols, numerous other QA/QC protocols must be employed to document the accuracy of the analytical portion of a waste evaluation program.

Health and Safety

Safety and health must also be considered when implementing a sampling plan. A comprehensive health and safety plan has three basic elements: (1) monitoring the health of field personnel; (2) routine safety procedures; and (3) emergency procedures.

Employees who perform field work, as well as those exposed to chemicals in the laboratory, should have a medical examination at the initiation of employment and routinely thereafter. This exam should preferably be performed and evaluated by medical doctors who specialize in industrial medicine. Some examples of parts of a medical examination that ought to be performed are: documentation of medical history; a standard physical exam; pulmonary functions screening; chest X-ray; EKG; urinalysis; and blood chemistry. These procedures are useful to: (1) document the quality of an employee's health at the time of matriculation; (2) ensure the maintenance of good health; and (3) detect early signs of bodily reactions to chemical exposures so they can be treated in a timely fashion. Unscheduled examinations should be performed in the event of an accident, illness, or exposure or suspected exposure to toxic materials.

Regarding safety procedures, personnel should be aware of the common routes of exposure to chemicals (i.e., inhalation, contact, and ingestion) and be instructed in the proper use of safety equipment, such as Draeger tube air samplers to detect air contamination, and in the proper use of protective clothing and respiratory equipment. Protocols should also be defined stating when safety equipment should be employed and designating safe areas where facilities are available for washing, drinking, and eating.

Even when the utmost care is taken, an emergency situation can occur as a result of an unanticipated explosion, electrical hazard, fall, or exposure to a hazardous substance. To minimize the impact of an emergency, field personnel should be aware of basic first aid and have immediate access to a first-aid kit. Phone numbers for both police and the nearest hospital should be obtained and kept by each team member before entering the site. Directions to the nearest hospital should also be obtained so that anyone suffering an injury can be transported quickly for treatment.

Chain of Custody

An essential part of any sampling/analytical scheme is ensuring the integrity of the sample from collection to data reporting. The possession and handling of samples should be traceable from the time of collection through analysis and final disposition. This documentation of the history of the sample is referred to as chain of custody.

Chain of custody is necessary if there is any possibility that the analytical data or conclusions based upon analytical data will be used in litigation. In

cases where litigation is not involved, many of the chain-of-custody procedures are still useful for routine control of sample flow. The components
of chain of custody — sample seals, a field logbook, chain-of-custody
record, and sample analysis request sheet — and the procedures for their use
are described in this section.

A sample is considered to be under a person's custody if it is (1) in a
person's physical possession, (2) in view of the person after taking possession, and (3) secured by that person so that no one can tamper with it, or
secured by that person in an area that is restricted to authorized personnel.
A person who has samples in custody must comply with the following
procedures.

(The material presented here briefly summarizes the major aspects of
chain of custody. The reader is referred to U.S. EPA NEIC Policies and Procedures, EPA -330/9/78/001-R [as revised 1/82], or other manual, as appropriate, for more information.)

Sample labels are necessary to prevent misidentification of samples.
Gummed paper labels or tags are adequate and should include at least the
following information:

- Sample number
- Name of collector
- Date and time of collection
- Place of collection

Labels should be affixed to sample containers prior to or at the time of
sampling and should be filled out at the time of collection.

Sample seals are used to detect unauthorized tampering of samples
following sample collection up to the time of analysis. Gummed paper
seals may be used for this purpose. The paper seal should include, minimally, the following information:

- Sample number (This number must be identical
  with the number on the sample label.)
- Name of collector
- Date and time of sampling
- Place of collection

The seal must be attached in such a way that it is necessary to break it
in order to open the sample container. Seals must be affixed to containers
before the samples leave the custody of sampling personnel.

All information pertinent to a field survey or sampling must be recorded
in a logbook. This should be bound, preferably with consecutively numbered pages that are 21.6 by 27.9 cm (8-1/2 by 11 in.). At a minimum, entries
in the logbook must include the following:

- Location of sampling point
- Name and address of field contact
- Producer of waste and address, if different from location
- Type of process producing waste (if known)
- Type of waste (e.g., sludge, wastewater)
- Suspected waste composition, including concentrations
- Number and volume of sample taken
- Purpose of sampling (e.g., surveillance, contract number)
- Description of sampling point and sampling methodology
- Date and time of collection
- Collector's sample identification number(s)
- Sample distribution and how transported (e.g., name of laboratory, UPS, Federal Express)
- References, such as maps or photographs of the sampling site
- Field observations
- Any field measurements made (e.g., pH, flammability, explosivity)
- Signatures of personnel responsible for observations

Sampling situations vary widely. No general rule can be given as to the extent of information that must be entered in the logbook. A good rule, however, is to record sufficient information so that anyone can reconstruct the sampling without reliance on the collector's memory. The logbook must be stored safely.

To establish the documentation necessary to trace sample possession from the time of collection, a chain-of-custody record should be filled out and should accompany every sample. This record becomes especially important if the sample is to be introduced as evidence in a court litigation. The record should contain, minimally, the following information:

- Sample number
- Signature of collector
- Date and time of collection
- Place and address of collection
- Waste type
- Signature of persons involved in the chain of possession
- Inclusive dates of possession

The sample analysis request sheet is intended to accompany the sample on delivery to the laboratory. The field portion of this form is completed by the person collecting the sample and should include most of the pertinent information noted in the logbook. The laboratory portion of this form is intended to be completed by laboratory personnel and to include, minimally:

- Name of person receiving the sample
- Laboratory sample number
- Date and time of sample receipt
- Sample allocation
- Analyses to be performed

The sample should be delivered to the laboratory for analysis as soon as practicable — usually within 1 or 2 days after sampling. The sample must be accompanied by the chain-of-custody record and by a sample analysis request sheet. The sample must be delivered to the person in the laboratory authorized to receive samples (often referred to as the sample custodian).

Any material that is identified in the DOT Hazardous Material Table (49 CFR §172.101) must be transported as prescribed in the table. All other hazardous waste samples must be transported as follows:

1. Collect sample in a 16-oz or smaller glass or polyethylene container with nonmetallic Teflon-lined screw cap. For liquids, allow sufficient air space (approximately 10% by volume) so that the container is not full at 54°C (130°F). If collecting a solid material, the container plus contents should not exceed 1 lb net weight. If sampling for volatile organic analysis, fill VOA container to septum but place the VOA container inside a 16-oz or smaller container so that the required air space may be provided. Large quantities, up to 3.785 liters (1 gal.), may be collected if the sample's flash point is 23 °C (75 °F) or higher. In this case, the flash point must be marked on the outside container (e.g., carton or cooler), and shipping papers should state that "Flash point is 73°F or higher."

2. Seal sample and place in a 4-mil-thick polyethylene bag, one sample per bag.

3. Place sealed bag inside a metal can with noncombustible, absorbent cushioning material (e.g., vermiculite or earth) to prevent breakage, one bag per can. Pressure-close the can and use clips, tape, or other positive means to hold the lid securely.

4. Mark the can with:
   Name and address of originator.
   "Flammable Liquid, N.O.S. UN 1993"
   (or, "Flammable Solid, N.O.S. UN 1325")

NOTE: UN numbers are now required in proper shipping names.

---

5. Place one or more metal cans in a strong outside container such as a picnic cooler or fiberboard box. Preservatives are not used for hazardous waste site samples.

6. Prepare for shipping: The words "Flammable Liquid, N.O.S. UN 1993" or "Flammable Solid, N.O.S. UN 1325"; "Cargo Aircraft Only" (if more than 1 qt. net per outside package); "Limited Quantity" or "Ltd. Qty."; "Laboratory Samples"; "Net Weight __" or "Net Volume __" (of hazardous contents) should be indicated on shipping papers and on the outside of the outside shipping container. The words "This Side Up" or "This End Up" should also be on container. Sign the shipper certification.

7. Stand by for possible carrier requests to open outside containers for inspection or to modify packaging. (It is wise to contact carrier before packing to ascertain local packaging requirements.) Remain in the departure area until the carrier vehicle (aircraft, truck, etc.) is on its way.

At the laboratory, a sample custodian should be assigned to receive the samples. Upon receipt of a sample, the custodian should inspect the condition of the sample and the sample seal, reconcile the information on the sample label and seal against that on the chain-of-custody record, assign a laboratory number, log in the sample in the laboratory logbook, and store it in a secured sample storage room or cabinet until it is assigned to an analyst for analysis.

The sample custodian should inspect the sample for any leakage from the container. A leaky container containing a multiphase sample should not be accepted for analysis. This sample will no longer be a representative sample. If the sample is contained in a plastic bottle and the container walls show that the sample is under pressure or releasing gases, the sample should be treated with caution because it may be explosive or release extremely poisonous gases. The custodian should examine whether the sample seal is intact or broken, because a broken seal may mean sample tampering and would make analysis results inadmissible as evidence in court. Any discrepancies between the information on the sample label and seal and the information that is on the chain-of-custody record and the sample analysis request sheet should be resolved before the sample is assigned for analysis. This effort might require communication with the sample collector. Results of the inspection should be noted on the sample analysis request sheet and on the laboratory sample logbook.

Incoming samples usually carry the inspector's or collector's identification numbers. To identify these samples further, the laboratory should assign its own identification numbers, which normally are given consecutively. Each sample should be marked with the assigned laboratory number. This number is correspondingly recorded on a laboratory sample log book along with the information describing the sample. The sample information is copied from the sample analysis request sheet and cross-checked against that on the sample label.

In most cases, the laboratory supervisor assigns the sample for analysis. The supervisor should review the information on the sample analysis request sheet, which now includes inspection notes recorded by the laboratory sample custodian. The technician assigned to analysis should record in the laboratory notebook the identifying information about the sample, the date of receipt, and other pertinent information. This record should also include the subsequent testing data and calculations. The sample may have to be split with other laboratories in order to obtain all the necessary analytical information. In this case, the same type of chain-of-custody procedures must be employed while the sample is being transported and at the other laboratory.

Once the sample has been received in the laboratory, the supervisor or his/her assignee is responsible for its care and custody. That person should be prepared to testify that the sample was in his/her possession or secured in the laboratory at all times, from the moment it was received from the custodian until the analyses were performed.

SAMPLE PLAN IMPLEMENTATION

Prior to implementing a sampling plan, it is often strategic to walk through the sampling plan mentally, starting with the preparation of equipment until the time when samples are received at the laboratory. This mental excursion should be in as much detail as can be imagined, because the small details are the ones most frequently overlooked. By employing this technique, items not included on the equipment list may be discovered, as well as any major oversight that could cause the sampling effort to fail. During this review of the sampling plan, an attempt should be made to anticipate what could go wrong. A solution to anticipated problems should be found, and, if necessary, materials needed for solving these problems should be added to the equipment list.

The remainder of this section discusses examples of sampling strategies for different situations that may be encountered.

Containers

Prior to discussing the sampling of containers, the term must be defined.

The term container, as used here, refers to receptacles that are designed for transporting materials, e.g., drums and other smaller receptacles, as opposed to stationary tanks. Weighted bottles, Coliwasas, drum thiefs, or triers are the sampling devices that are chosen for the sampling of containers.

The sampling strategy for containers varies according to (1) the number of containers to be sampled and (2) access to the containers. Ideally, if the waste is contained in several containers, every container will be sampled. If this is not possible due to the large number of containers or to cost factors, a subset of individual containers must be randomly selected for sampling. This can be done by assigning each container a number and then randomly choosing a set of numbers for sampling.

Access to a container will affect the number of samples that can be taken from the container and the location within the container from which samples can be taken. Ideally, several samples should be taken from locations displaced both vertically and horizontally throughout the waste. The number of samples required for reliable sampling will vary depending on the distribution of the waste components in the container. At a minimum with an unknown waste, a sufficient number and distribution of samples should be taken to address any possible vertical anomalies in the waste. This is because contained wastes have a much greater tendency to be nonrandomly heterogeneous in a vertical rather than a horizontal direction due to (1) settling of solids and the denser phases of liquids, and (2) variation in the content of the waste as it enters the container. Bags, paper drums, and open-headed steel drums (of which the entire top can be removed) generally do not restrict access to the waste and therefore do not limit sampling.

When access to a container is unlimited, a useful strategy for obtaining a representative set of samples is a three-dimensional simple random sampling strategy in which the container is divided by constructing an imaginary three-dimensional grid, as follows. First, the top surface of the waste is divided into a grid whose sections either approximate the size of the sampling device or are larger than the sampling device if the container is large. (Cylindrical containers can be divided into imaginary concentric circles, which are then further divided into grids of equal size.) Each section is assigned a number. The height of the container is then divided into imaginary levels that are at least as large as the vertical space required by the chosen sampling device. These imaginary levels are then assigned numbers. Specific levels and grid locations are then selected for sampling using a random-number table or random-number generator.

Another appropriate sampling approach is the two-dimensional simple random sampling strategy, which can usually yield a more precise sampling when fewer samples are collected. This strategy involves (1) dividing the top surface of the waste into an imaginary grid as in the three-dimensional strategy, (2) selecting grid sections for sampling using random-number tables or random-number generators, and (3) sampling each selected grid point in a vertical manner along the entire length from top to bottom using a sampling device such as a drum thief or Coliwasa.

Some containers, such as drums with bung openings, limit access to the contained waste and restrict sampling to a single vertical plane. Samples taken in this manner can be considered representative of the entire container only if the waste is known to be homogeneous or if no horizontal stratification has ocurred. Precautions must be taken when sampling any type of steel drum because the drum may explode or expel gases and/or pressurized liquids. An EPA/NEIC manual, "Safety Manual for Hazardous Waste Site Investigation," addresses these safety precautions.

Tanks

Tanks are essentially large containers. The considerations involved in sampling tanks are therefore similar to those for sampling containers. As with containers, the goal of sampling tanks is to acquire a sufficient number of samples from different locations within the waste to provide analytical data that are representative of the entire tank contents.

The accessibility of the tank contents will affect the sampling methodology. If the tank is an open one, allowing unrestricted access, then usually a representative set of samples is best obtained using the three-dimensional simple random sampling strategy, as described for containers. This strategy involves dividing the tank contents into an imaginary three-dimensional grid. As a first step, the top surface of the waste is divided into a grid whose sections either approximate the size of the sampling device or are larger than the sampling device if the tank is large. (Cylindrical tanks can be divided into imaginary concentric circles, which are then further divided into grids of equal size.) Each section is assigned a number. The height of the tank is then divided into imaginary levels that are at least as large as vertical space required by the chosen sampling device. These imaginary levels are assigned numbers. Specific levels and grid locations are then selected for sampling using a random-number table or random-number generator.

A less comprehensive sampling approach may be appropriate if information regarding the distribution of waste components is known or assumed (e.g., if vertical compositing will yield a representative sample). In such cases, a two-dimensional simple random sampling strategy may be appropriate. In this strategy, the top surface of the waste is divided into an imaginary grid; grid sections are selected using random-number tables or random-number generators; and each selected grid point is then sampled in a vertical manner along the entire length from top to bottom using a sampling device such as weighted bottle, a drum thief, or Coliwasa. If the waste is known to consist of two or more discrete strata, a more precise representation of the tank contents can be obtained by using a stratified random sampling strategy, i.e., by sampling each stratum separately using the two- or three-dimensional simple random sampling strategy.

Some tanks permit only limited access to their contents, which restricts the locations within the tank from which samples can be taken. If sampling is restricted, the sampling strategy must, at a minimum, take sufficient samples to address the potential vertical anomalies in the waste in order to be considered representative. This is because contained wastes tend to display vertical, rather than horizontal, nonrandom heterogeneity due to settling of suspended solids or denser liquid phases. If access restricts sampling to a portion of the tank contents (e.g., in an open tank, the size of the tank may restrict sampling to the perimeter of the tank; in a closed tank, the only access to the waste may be through inspection ports), then the resulting analytical data will be deemed representative only of the accessed area, not of the entire tank contents unless the tank contents are known to be homogeneous.

If a limited access tank is to be sampled, and little is known about the distribution of components within the waste, a set of samples that is representative of the entire tank contents can be obtained by taking a series of samples as the tank contents are being drained. This should be done in a simple random manner by estimating how long it will take to drain the tank and then randomly selecting times during drainage for sampling.

The most appropriate type of sampling device for tanks depends on the tank parameters. In general, subsurface samplers (i.e., pond samplers) are used for shallow tanks, and weighted bottles are usually employed for tanks deeper than 5 ft. Dippers are useful for sampling pipe effluents.

Waste Piles

In waste piles, the accessibility of waste for sampling is usually a function of pile size, a key factor in the design of a sampling strategy for a waste pile. Ideally, piles containing unknown wastes should be sampled using a three-dimensional simple random sampling strategy. This strategy can be employed only if all points within the pile can be accessed. In such cases, the pile should be divided into a three-dimensional grid system, the grid sections assigned numbers, and the sampling points then chosen using random-number tables or random-number generators.

If sampling is limited to certain portions of the pile, then the collected sample will be representative only of those portions, unless the waste is known to be homogeneous.

In cases where the size of a pile impedes access to the waste, a set of samples that are representative of the entire pile can be obtained with a minimum of effort by scheduling sampling to coincide with pile removal. The number of truck loads needed to remove the pile should be estimated and the truckloads randomly chosen for sampling.

The sampling devices most commonly used for small piles are thiefs, triers, and shovels. Excavation equipment, such as backhoes, can be useful for sampling medium-sized piles.

Landfills and Lagoons

Landfills contain primarily solid waste, whereas lagooned waste may range from liquids to dried sludge residues. Lagooned waste that is either liquid or semisolid is often best sampled using the methods recommended for large tanks. Usually, solid wastes contained in a landfill or lagoon are best sampled using the three-dimensional random sampling strategy.

The three-dimensional random sampling strategy involves establishing an imaginary three-dimensional grid of sampling points in the waste and then using random-number tables or random-number generators to select points for sampling. In the case of landfills and lagoons, the grid is established using a survey or map of the area. The map is divided into two two-dimensional grids with sections of equal size. These sections are then assigned numbers sequentially.

Next, the depth to which sampling will take place is determined and subdivided into equal levels, which are also sequentially numbered. (The lowest sampling depth will vary from landfill to landfill. Usually, sampling extends to the interface of the fill and the natural soils. If soil contamination is suspected, sampling may extend into the natural soil.) The horizontal and vertical sampling coordinates are then selected using random-number tables or random-number generators. If some information is known about the nature of the waste, then a modified three-dimensional strategy may be more appropriate. For example, if the landfill consists of several cells, a more precise measurement may be obtained by considering each cell as a stratum and employing a stratified three-dimensional random sampling strategy.

Hollow-stem augers combined with split-spoon samplers are frequently appropriate for sampling landfills. Water-driven or water-rinsed coring equipment should not be used for sampling because the water can rinse chemical components from the sample. Excavation equipment, such as backhoes, may be useful in obtaining samples at various depths; the resulting holes may be useful for viewing and recording the contents of the landfill.

SAMPLE COMPOSITING

The compositing of samples, usually done for cost-saving reasons, involves the combining of a number of samples or aliquots of a number of samples collected from the same waste. The disadvantage of sample compositing is the loss of concentration variance data, whereas the advantage is that, for a given analytical cost, a more representative (i.e., more accurate) sample is obtained.

It is usually most expedient and cost effective to collect component samples in the field and to composite aliquots of each sample later in the laboratory. Then, if after reviewing the data any questions arise, the samples can be recomposited in a different combination, or each component sample can be analyzed separately to determine better the variation of waste composition over time and space, or to determine better the precision of an average number. The fact that this recompositing of samples can occur without the need to resample often results in a substantial cost savings.

To ensure that recompositing can be done at a later date, it is essential to collect enough sample volume in the field so that, under normal circumstances, enough component sample will remain following compositing to allow for a different compositing scheme or even for an analysis of the component samples themselves.

The actual compositing of samples requires the homogenization of all component samples to ensure that a representative subsample is aliquoted. The homogenization procedure, and the containers and equipment used for compositing, will vary according to the type of waste being composited and the parameters to be measured. Likewise, the composite sample itself will be homogenized prior to the subsampling of analytical aliquots.

Appendix E

Economic Analysis of Pollution
Prevention Projects

The following text was extracted from from the U.S. Environmental Protection Agency's Facility Pollution Prevention Guide published by the Office of Research and Development, Washington, D.C. 20460. May 1992.

Although businesses may invest in pollution prevention because it is the right thing to do or because it enhances their public image, the viability of many prevention investments rests on sound economic analyses. In essence, companies will not invest in a pollution prevention project unless that project successfully competes with alternative investments. The purpose of this chapter is to explain the basic elements of an adequate cost accounting system and how to conduct a comprehensive economic assessment of investment options.

TOTAL COST ASSESSMENT

In recent years industry and the EPA have begun to learn a great deal more about full evaluation of prevention-oriented investments. In the first place, we have learned that business accounting systems do not usually track environmental costs so they can be allocated to the particular production units that created those wastes. Without this sort of information, companies tend to lump environmental costs together in a single overhead account or simply add them to other budget line items where they cannot be disaggregated easily. As a result, companies do not have the ability to identify those parts of their operations that cause the greatest environmental expenditures or the products that are most responsible for waste production. This chapter provides some guidance on how accounting systems can be set up to capture this useful information better.

It has also become apparent that economic assessments typically used for investment analysis may not be adequate for pollution prevention projects. For example, traditional analysis methods do not adequately address the fact that many pollution prevention measures will benefit a larger number of production areas than do most other kinds of capital investment. Second, they do not usually account for the full range of environmental expenses companies often incur. Third, they usually do not accommodate

201

a sufficiently long time horizon to allow full evaluation of the benefits of many pollution prevention projects. Finally, they provide no mechanism for dealing with the probabilistic nature of pollution prevention benefits, many of which cannot be estimated with a high degree of certainty. This chapter provides guidance on how to overcome these problems as well.

In recognition of opportunities to accelerate pollution prevention, the U.S. EPA has funded several studies to demonstrate how economic assessments and accounting systems can be modified to improve the competitiveness of prevention-oriented investments. EPA calls this analysis Total Cost Assessment (TCA). There are four elements of Total Cost Assessment: expanded cost inventory, extended time horizon, use of long-term financial indicators, and direct allocation of costs to processes and products. The first three apply to feasibility assessment, while the fourth applies to cost accounting. Together these four elements will help you to demonstrate the true costs of pollution to your firm as well as the net benefits of prevention. In addition, they help you show how prevention-oriented investments compete with company-defined standards of profitability. In sum, TCA provides substantial benefits for pre-implementation feasibility assessments (see Chapter 2 on preliminary assessments and Chapter 3 on feasibility analysis) and for post-implementation project evaluation (see Chapter 4 on measuring progress).

The remainder of this chapter summarizes the essential characteristics of TCA. Much of the information is drawn from a report recently prepared for the U.S. EPA by Tellus Institute. (See Appendix G for the full citation.) The Tellus report addresses TCA methodology in much greater detail than can be provided here and provides examples of specific applications from the pulp and paper industry. The report also includes an extensive bibliography on applying TCA to pollution prevention projects. In a separate but related study for the New Jersey Department of Environmental Protection, Tellus analyzed TCA as it applies to smaller and more varied industrial facilities. A copy of this report can be obtained from the N.J. Department of Environmental Protection.

EXPANDED COST INVENTORY

TCA includes not only the direct cost factors that are part of most project cost analyses, but also indirect costs, many of which do not apply to other types of projects. Besides direct and indirect costs, TCA includes cost factors related to liability and to certain "less-tangible" benefits.

TCA is a flexible tool that can be adapted to your specific needs and circumstances. A full-blown TCA will make more sense for some businesses than for others. In either case it is important to remember that TCA can happen incrementally by gradually bringing each of its elements to the investment evaluation process. For example, while it may be quite easy to obtain information on direct costs, you may have more trouble estimating some of the future liabilities and less tangible costs. Perhaps your first effort should incorporate all direct costs and as many indirect costs as possible. Then you might add those costs that are more difficult to estimate as increments to the initial analysis, thereby highlighting to management both their uncertainty and their importance.

DIRECT COSTS

For most capital investments, the direct cost factors are the only ones considered when project costs are being estimated. For pollution prevention projects, this category may be a net cost, even though a number of the components of the calculation will represent savings. Therefore, confining the cost analysis to direct costs may lead to the incorrect conclusion that pollution prevention is not sound business investment.

INDIRECT COSTS

For pollution prevention projects, unlike more familiar capital investments, indirect costs are likely to represent a significant net savings. Administrative costs, regulatory compliance costs (such as permitting, recordkeeping, reporting, sampling, preparedness, closure/post-closure assurance), insurance costs, and on-site waste management and pollution control equipment operation costs can be significant. They are considered hidden in the sense that they are either allocated to overhead rather than their source (production process or product) or are altogether omitted from the project financial analysis. A necessary first step in including these costs in an economic analysis is to estimate and allocate them to their source. See the section below on Direct Cost Allocation for several ways to accomplish this.

LIABILITY COSTS

Reduced liability associated with pollution prevention investments may also offer significant net savings to your company. Potential reductions in penalties, fines, cleanup costs, and personal injury and damage claims can make prevention investments more profitable, particularly in the long run.

In many instances, estimating and allocating future liability costs is subject to a high degree of uncertainty. It may, for example, be difficult to estimate liabilities from actions beyond your control, such as an accidental spill by a waste hauler. It may also be difficult to estimate future penalties and fines that might arise from noncompliance with regulatory standards that do not yet exist. Similarly, personal injury and property damage claims that may result from consumer misuse, from disposal of waste later classified as hazardous, or from claims of accidental release of hazardous waste after disposal are difficult to estimate. Allocation of future liabilities to the products or production processes also presents practical difficulties in a cost assessment. Uncertainty, therefore, is a significant aspect of a cost assessment and one that top management may be unaccustomed to or unwilling to accept.

Some firms have nevertheless found alternative ways to adds liability costs in project analysis. For example, in the narrative accompanying a profitability calculation, you could include a calculated estimate of liability reduction, cite a penalty or settlement that may be avoided (based on a claim against a similar company using a similar process), or qualitatively indicate without attaching dollar value the reduced liability risk associated with the pollution prevention project. Alternatively, some firms have chosen to loosen the financial performance requirements of their projects to account for liability reductions. For example, the required payback period can be lengthened from three to four years, or the required internal rate of return can be lowered from 15 to 10 percent. (See the U.S. EPA's *Pollution Prevention Benefits Manual,* Phase II, as referenced in Appendix G, for suggestions on formulas that may be useful for incorporating future liabilities into the cost analysis.)

Less-Tangible Benefits

A pollution prevention project may also deliver substantial benefits from an improved product and company image or from improved employee health. These benefits, listed in the cost allocation section of this chapter, remain largely unexamined in environmental investment decisions. Although they are often difficult to measure, they should be incorporated into the assessment whenever feasible. At the very least, they should be highlighted for managers after presenting the more easily quantifiable and allocatable costs.

Consider several examples. When a pollution prevention investment improves product performance to the point that the new product can be differentiated from its competition, market share may increase. Even conservative estimates of this increase can incrementally improve the payback from the pollution prevention investment. Companies similarly recognize that the development and marketing of so-called "green products" appeals to consumers and increasingly appeals to intermediate purchasers who are interested in incorporating "green" puts into their products. Again, estimates of potential increases in sales can be added to the analysis. At the very least, the improved profitability from adding these less-tangible benefits to the analysis should be presented to management alongside the more easily estimated costs and benefits. Other less tangible benefits may be more difficult to quantify, but should nevertheless be brought to management's attention. For example, reduced health maintenance costs, avoided future regulatory costs, and improved relationships with regulators potentially affect the bottom line of the assessment.

In time, as the movement toward green products and companies grows, as workers come to expect safer working environments, and as companies move away from simply reacting to regulations and toward anticipating and addressing the environmental impacts of their processes and products, the less tangible aspects of pollution prevention investments will become more apparent.

EXPANDED TIME HORIZON

Since many of the liability and less-tangible benefits of pollution prevention will occur over a long period of time, it is important that an economic assessment look at a long time frame, not the three to five years typically used for other types of projects. Of course, increasing the time frame increases the uncertainty of the cost factors used in the analysis.

LONG-TERM FINANCIAL INDICATORS

When making pollution prevention decisions, select long-term financial indicators that account for:
- all cash flows during the project
- the time value of money.

Three commonly used financial indicators meet these criteria: Net Present Value (NPV) of an investment, Internal Rate of Return (IRR), and Profitability Index (PI). Another commonly used indicator, the Payback Period, does not meet the two criteria mentioned above and should not be used.

Discussions on using these and other indicators will be found in economic analysis texts.

DIRECT ALLOCATION OF COSTS

Few companies allocate environmental costs to the products and processes that produce these costs. Without direct allocation, businesses tend to lump these expenses into a single overhead account or simply add them to other budget line items where they cannot be disaggregated easily. The result is an accounting system that is incapable of (1) identifying the products or processes most responsible for environmental costs, (2) targeting prevention opportunity assessments and prevention investments to the high environmental cost products and processes, and (3) tracking the financial savings of a chosen prevention investment. TCA will help you remedy each of these deficiencies.

Like much of the TCA method, implementation of direct cost allocation should be flexible and tailored to the specific needs of your company. To help you evaluate the options available to you, the discussion below introduces three ways of thinking about allocating your costs: single pooling, multiple pooling, and service centers. The discussion is meant as general guidance and explains some of the advantages and disadvantages of each approach. Please see other EPA publications (such as those listed in Appendix G), general accounting texts, and financial specialists for more detail.

SINGLE POOL CONCEPT

With the single pool method, the company distributes the benefits and costs of pollution prevention across all of its products or services. A general overhead or administrative cost is included in all transactions.

Advantages. This is the easiest accounting method to put into use. All pollution costs are included in the general or administrative overhead costs that most companies already have, even though they may not be itemized as pollution costs. It may therefore not be a change in accounting methods but rather an adjustment in the overhead rate. No detailed accounting or tracking of goods is needed. Little additional administrative burden is incurred to report the benefits of pollution prevention.

Disadvantages. If the company has a diverse product or service line, pollution costs may be covered from products or services that do not contribute to the pollution. This has the effect of inflating the costs of those products or services unnecessarily. It also obscures the benefits of pollution prevention to the people who have the opportunity to make it successful; the line manager will not see the effect of preventing or failing to prevent pollution in his area of responsibility.

MULTIPLE POOL CONCEPT

The next level of detail in the accounting process is the multiple pool concept, wherein pollution prevention benefits or costs are recovered by the department or other operating unit level.

Advantages. This approach ties the cost of pollution more closely to the responsible activity and to the people responsible for daily implementation. It is also easy to apply within an accounting system that is already set up for departmentalized accounting.

Disadvantages. A disparity may still exist between responsible activities and the cost of pollution. For example, consider a department that produces parts for many outside companies. Some customers need standard parts, while others require some special preparation of the parts. This special preparation produces pollution. [Is it] reasonable to allocate the benefit or cost for this pollution prevention project across all of the parts produced?

SERVICE CENTER CONCEPT

A much more detailed level of accounting is the service center concept. Here, the benefits or costs of pollution prevention are allocated to only those activities that are directly responsible.

Advantages. Pollution costs are accurately tied to the generator. Theoretically, this is the most equitable to all products or services produced. Pollution costs can be identified as direct costs on the appropriate contracts and not buried in the indirect costs, affecting competitiveness on other contracts. Pollution costs are more accurately identified, monitored and managed. The direct benefits of pollution prevention are more easily identified and emphasized at the operational level.

Disadvantages. Considerable effort may be required to track each product, service, job, or contract and to recover the applicable pollution surcharges. Added administrative costs may be incurred to implement and maintain the system. It may be difficult to identify the costs of pollution when pricing and order or bidding on a new contract. It may be difficult to identify responsible activities under certain circumstances such as laboratory services where many small volumes of waste are generated on a seemingly continual basis.

SUMMARY

Environmental costs have been rising steadily for many years now. Initially, these costs did not seem to have a major impact on production. For this reason, most companies simply added these costs to an aggregate overhead account, if they tracked them at all. The tendency of companies to treat environmental costs as overhead and to ignore many of the direct, indirect, and less-tangible environmental costs (including future liability) in their investment decisions has driven the development of TCA.

Expanding your cost inventory pulls into your assessments a much wider array of environmental costs and benefits. Extending the time horizon, even slightly, can improve the profitability of prevention investments substantially, since these investments tend to have somewhat longer payback schedules. Choosing long-term financial indicators, which consistently provide managers with accurate and comparable project financial assessments, allows prevention oriented investments to compete successfully with other investment options. Finally, directly allocating costs to processes and products enhances your ability to target prevention investments to high environmental cost areas, routinely provides the information needed to do TCA analysis, and allows managers to track the success of prevention investments. Overall, the TCA method is a flexible tool, to be applied incrementally, as your company's needs dictate.

Abbreviations and Acronyms

33/50 Project	voluntary EPA program calling for reductions (33% by 1992 and 50% by 1995) in releases and off site transfers of 17 industrial chemicals/chemical categories
t1/2	time required for a 50% reduction in mass of the original substance
ACGIH	American Conference of Governmental Industrial Hygienists
AICHE	American Institute of Chemical Engineers
ASTM	American Society for Testing and Materials
ATSDR	Agency for Toxic Substances and Disease Registry
CAS	Chemical Abstract Service
CFC	Chlorofluorocarbon
CERCLA	Comprehensive Environmental Response, Compensation and Liability Act
CFR	Code of Federal Regulations
CNS	Central Nervous System
CO	Carbon Monoxide
CO2	Carbon Dioxide
DCM	Dichloromethane
DF	Designated Facility
DOT	Department of Transportation
DNA	Deoxyribonucleic acid
EPCRA	Emergency Planning and Community Right-to-Know Act
EPA	Environmental Protection Agency
FOIA	Freedom of Information Act
LEC	Lowest Effect Concentration
HSWA	Hazardous and Solid Waste Amendments
IRIS	Integrated Risk Information System
MBtu	Mega (10^6)-British Thermal Units
MEK	Methyl Ethyl Ketone
MSDS	Material Safety Data Sheet
NAS	National Academy of Sciences
NFPA	National Fire Protection Association
N.O.S.	Not Otherwise Specified

NIOSH	National Institute for Occupational Safety and Health
NTIS	National Technical Information Service
NTP	National Toxicology Program
OSHA	Occupational Safety and Health Administration
ppm	parts per million
ppb	parts per billion
QA/QC	Quality Assurance/Quality Control
RCRA	Resource Conservation and Recovery Act
RTECS	Registry of Toxic Effects of Chemical Substances
SARA	Superfund Amendments and Reauthorization Act
STEL	Short Term Exposure Limit
TCA	Total Cost Assessment
TCA	1,1,1-Trichloroethane
TCE	Trichloroethylene
TLV	Threshold Limit Value
TNT	Trinitrotoluene
TCLP	Toxicity Characteristic Leaching Procedure
TSCA	Toxic Substances Control Act
TSDF	Treatment, Storage or Disposal Facility
TWA	Time Weighted Average
UN	United Nations
WWT	Wastewater Treatment

Index

X